BORO RICE CULTIVATION AND PRODUCTIVITY

BORO RICE CULTIVATION AND PRODUCTIVITY

S.K. Jha

ABHIJEET PUBLICATIONS
DELHI 110094

ABHIJEET PUBLICATIONS
2/46 Tukhmeerpur Extension
Delhi-110 094
Phone: 011-22960492, 65698474
E-mail: abhijeetpublication@gmail.com

BORO RICE CULTIVATION AND PRODUCTIVITY

First Published 2010

ISBN 978-93-80031-81-1

PRINTED IN INDIA

Published by J.K. Singh for Abhijeet Publications, Delhi-110 094, Lasertypeset by Gaurav Graphics, New Delhi and Printed at Himanshu Printers, Delhi.

Preface

Training is a critical input for the rapid transfer of agricultural technology to the user. It aims at helping each individual by way of increased knowledge improved skill and changed attitude enabling him to perform his job with better consequences. There is already a wide gap between the transferable technology and transferred technology as far as the main beneficiaries are concerned. The farmers of lower socio-economic status and of less education having less knowledge about modern agricultural technologies can be well equipped through effective training programmes. Boro rice has a high production potential of 7.5-8.9 tonnes/ha with the possibilities of further increase under better management. Its cultivation is increasing not only in Bihar but in other states also. The strategy for transfer of technology of new innovation of boro rice over the year has not proved to be successful because of gaps in the field application of technology. Farmers need and preferences have remained still un-explored. It needed the training programme tailored to the needs and problems of farmers to be organized to improve their agricultural productivity.

Objective of the Study

The present study has planned to convert the available boro rice technology into production accomplishments with the following objectives:

1. To identify agro-economic, socio-psychological and communicational characteristics of boro rice growing farmers.
2. To assess the training needs in respect of scientific boro rice production as perceived by farmers.
3. To find out the association of the training needs with certain selected agro-economic, socio-psychological and communicational variables.
4. To seek the opinion of farmers in respect of different components of training programme.
5. To develop training strategy for different categories of farmers for increasing boro rice production.

Methodology

Madhubani district of North Bihar was purposely selected for this study. Two blocks namely Madhepur and Lakhnaur were purposively selected. A total of 200 boro rice growing farmers belonging to marginal, small and medium categories were selected for the study on the basis of probability proportionate to the number of farm families under each category.

Summary and Findings

In case of marginal farmers, the highest technological gap was observed in the use of improved seed and their treatment followed by fertilizer application, sowing methods and irrigation. The highest gap was found in the practice of "fertilizer application" and irrigation by small as well as medium farmers. Improved cultural practices emerged as the top most priority training area as there was common consciousness in all the three categories of farmers in having training in this area as most needed. The finding reveals that use of integrated nutrient management and water management were preferred as a priority area by the marginal, small and medium farmers respectively. In the sub-areas of training, transplanting of boro rice seedling, identification of HYV rice, time of irrigation, method of storage and weed control by manual, chemical and mechanical methods have been given top priority by all the three categories of farmers.

The variables cropping intensity and economic motivation were found to be positively and significantly associated with training need in all the three categories of farmers. The variables farm size, education of the farmer and utilization of personal cosmopolite were positively and significantly associated with training need of marginal farmers in boro rice technology. Technological gap and utilization of mass media were found to be positively and significantly associated with training need of small farmers. Variables farm mechanization, risk orientation, market orientation, utilization of mass media and utilization of personal cosmopolite sources were found to have positive and significant association with training need of medium farmers. Variables farm size, education, of the farmers, economic motivation and cropping intensity emerged as significant contributors towards identifying the training need of marginal farmers. Only three variables namely economic motivation, cropping intensity and utilization of mass media emerged as powerful predictors in case of small farmers. Variables market orientation, utilization of mass media and economic motivation emerged as powerful predictors in case of medium farmers in boro rice technology. The variables farm size, economic motivation and market orientation were the deciding factors affecting the training need as these variables yielded the highest substantial direct effect on training need of marginal, small and medium farmers respectively.

Organization of training programme in the village during slack season was most preferred by small framers. However, medium farmers preferred the university/ college campus and the time of training before the commencement of crop season. All the three categories of farmers showed their desire for organization of training programme with a group of 20 to 25 farmers to be imparted by the boro rice scientists (university) and also their interaction with farmers at regular intervals after participating in the training programme. In case of single method of training question-answer method was ranked first by the marginal farmers whereas small and medium farmers gave top priority to training by demonstration at the field level. Under combination of three methods, lecture + method demonstration + practice by trainees was ranked first by marginal farmers, whereas small and medium farmers gave top priority for the training methods as lecture + method demonstration + published material. Under combination of four methods, lecture + method demonstration + practice by trainees + published material was ranked first by marginal farmers whereas small and medium farmers gave top priority for training methods as method demonstration + question-answer + tour + published material. Under combination of five methods, audio visual aids + practice by trainees + question-answer + published material + tour was ranked first by marginal farmers. Whereas, small and medium farmers gave top priority for training methods as lecture + method demonstration + practice by trainees + tour + publication.

This work would have not been possible without the support of friends and well wishers. The words are inadequate to express my gratitude to them. First of all, I am grateful to my guide Dr. K. Mishra, Additional Director Academic, Lalit Narayan Mishra College of Business Management & Dean, Faculty of Management, BRA Bihar University Muzaffarpur for his precious and imposing guidance and constructive criticism during the course of investigation and preparation of the dissertation.

I express my profound sense of gratitude to Dr. B.C. Chaudhary, Director Research, RAU and Dr. Ramanand Yadav, University Professor, Cum-Chief Scientist, RAU, Pusa, Bihar for their constant help and valuable suggestions at different stages of report writing.

I am highly grateful to Dr. N.K. Chaudhary, University Professor cum Chief Scientist, Dr. D.K. Roy, Assistant Professor cum Jr. Scientist, Department of Agronomy and Dr. N.K. Singh, Senior Rice Breeder, RAU, Pusa for their valuable suggestions during the course of investigation.

I am very much thankful to all the faculty members of department of Agricultural Economics for their kind cooperation.

I offer my sincere gratitude to my juniors who were associated directly or indirectly during the entire course of this investigation.

Within the family I owe my hearty appreciation, affection and loving gratefulness to each and every member, who not only encouraged me to carry out and complete this thesis, but also provided me with congenial environment to bring this work to fruition. Particularly I can not forget Nila Jha my wife, Madan Mohan Jha my son, Dharmendra Kumar Thakur my son-in-law, Madhuri Thakur my daughter and Mamta Jha my daughter-in-law. Words are inadequate to express my feeling and gratitude towards them.

S. K. Jha

Contents

1

Introduction

Boro rice is globally emerging a top ranking cereal in potential grain productivity and is third after wheat and maize, sharing about one-third of total cereal production. It is now an important ingredient of food, feed and a large number of industrial products. It has acquired a dominant role in the farming sector and in the macro-ecönomy of the Asian region. In Bihar, boro rice is not a traditional cereal grown mainly for human food purposes, though recently its industrial uses are steadily increasing. Boro word is derived from Sanskrit word "BOROB" refer to special rice cultivation in water logged low-lying area during November to May. Boro rice and wheat are grown in the same season. Boro rice is an alternative crop where wheat can not be grown due to excess of water in the field. The area under Boro rice has increased significantly in India. Boro rice is cultivated in 1.89 lakh hectare in India year after year. There is vast scope for its expansion especially in other areas of U.P., Bihar and Assam.

North-Eastern Bihar is highly flood affected area. There is need to improve the productivity of rice especially of rainfed lowland affected by flash flood and deep water ecosystem facing water logging problems. Boro rice cultivation may prove to be boon for augmenting the productivity of deepwater ecosystem. Madhubani is the most flood-affected district and 18 rivers and rivulets pass through this district from north to south. The major portion of the land lies between Kamala Balan and Tiljuga rivers which are known as "Sorrow of Bihar" as the Huang-ho in China. The eastern part across the Kosi river faces flash floods whereas, the western part faces water logging situation. The middle portion which is higher than the eastern and western part of Madhubani is confronted with a number of problems viz., flood, drought, water logging, non-consolidation of land, lack of irrigation facility, depleted roads, perpetual water logging, non-supply of inputs, micronutrient deficiencies, etc. Rainfed lowland and deepwater ecosystem are more prevalent and occupy > 75% of gross cropped area. Several bowl shaped depression locally known 'chaur' are found here where water stagnation takes place during June-July to February-march and rice can be neither grown in kharif season nor wheat in rabi season. A typical 'chaur' has higher depth in middle and shallow at the periphery. In such situation boro rice technology, recently has been introduced and area expansion took place tremendously. Due to low productivity of rainfed lowland and deepwater ecosystem and lack of agribased allied professions, migration of labor in huge number is taking place and over crusading the other cities. The most achievement of the independent India is the transformation of the country from chronic deficit to food self-reliant nation. Major contribution to the food production gains has come from rice (42%). India is the seventh largest country in the world by area, with 329 million hectares (M ha). It is also the second-most populous country (1 billion people). Demographers indicate that by 2012 India's population will reach 1.2 billion. Rice (*Oryza sativa* L.) has supported a greater number of people for a longer period of time than any other crop since it was domesticated between 8,000 to 10,000 years ago (Greenland, 1997). At present, rice is the staple food for more people than wheat, and 90 percent of total rice production is grown and consumed in Asia.

1.1 What is Boro Rice?

Boro rice is a winter irrigated rice crop with a high productivity potential. The crop is affected by low temperature at seedling stage and early vegetative stage during the period November to January. Low temperature sometimes kills the plants or prolongs growth duration. As a result of a prolonged growth period, varieties normally flower in April-May, the summer months and are often affected by drought and very high temperatures at the reproductive phase. Boro rice cultivation is a special system of rice cultivation in water logged low-lying or medium land areas with irrigation during November to May. The boro rice system takes advantage of residual moisture in lands after the harvest of Kharif paddy; longer moisture retention capacity of the soil; and surface water stored in nearby low-lying ditches, areas adjoining canals, roads, chaur lands, etc.

Ample possibilities for increasing the boro rice area exist provided suitable varieties and matching cultural practices are developed. There is need to develop varieties with tolerance for low temperature during seedling and vegetative phases. In the eastern and north-eastern Himalayan areas of India, boro rice is grown in about 2.5 million ha, the highest area of 1 million ha in West Bengal. At present boro rice is occupying an area of nearly 0.18 m.ha in Assam, 0.1 m.ha in Bihar and 0.4 m.ha in Tripura. Short duration varieties maturing the latest in mid-April while maintaining higher yield potential will require less inputs, particularly irrigation water. Therefore, boro rice cultivation on sound scientific basis by screening of "miracle" varieties along with new agricultural technology promises great potential in solving the puzzle of utilizing the unproductive low and thereby increasing rice yield.

Traditionally boro rice was cultivated in river basin deltas where water accumulated during monsoon and could not be drained out in the winter months. Boro rice is often grown in low lying deepwater rice fields to utilize residual water in the soil. It is often subjected to low temperature stress at seedling stage and early vegetative stage while hot weather or water stress at reproductive phase but major problem is severe cold injury at the seedling stage which is caused by low air temperature as well as low water temperature (Nanda and Sheshu, 1979). Physiological and biological changes are closely related with cold tolerance (Maruyama et al., 1990). In north-east India early sowing is practiced to avoid damage due to cold. During winter, seedlings show varied effect of cold damage on leaves which manifest in severe yellowing, infestation of brown spot and drying. The seedlings remain stunted during December-January and onwards transplanting is done. For expanding boro rice area, massive investment for tapping ground water, conservation of residual moisture, provision for live saving irrigation are need.

1.2 Why Boro Rice?

Due to change in environment with special reference to rainfall pattern and amount. These wet season aman rice crop face the problems of submergence at early stage due to flood and drought at panicle initiation stage and such adverse situation was prevailing during last years due to failure of Hathia rain. Such extremely diverse situation compelled to farmers to search an alternative for their substinance and livelihood which is mainly based on rice crop cultivation. The boro rice cultivation system emerges from such crucial situation and now it has become a boon for victims of natural vagaries prone areas. Vast area of rainfed lowland and deepwater ecologies remain fellow after harvest of monsoon rice in eastern India, inspite of the fact that there is an unlimited and annually rechargeable source of water underneath. Learning from the experiences of West Bengal, where 1 m/ha have been brought under boro (winter) rice, Bihar, Assam and Orissa with similar underutilized potential of ground water must take up similar measures. Investment in the development of minor irrigation (ground water utilization) in Bihar can bring a million more hectare under boro rice the yield of rice in Bihar (Siddiq, 1996). Selvaraju (1998) reported that the yield gap in rice in Bihar is 84 percent, which is the highest in comparison to other state, if we keep only 5.0 tonnes per hectare as the maximum attainable potential yield. Three aspect have to be considered for raising the global production of rice: Firstly raising the "yield ceiling". Secondly, narrowing the "yield gap". And, thirdly, while doing both of these, to maintain the "sustainability of production". The double rice cropping pattern in deepwater and floating rice areas will expand when researcher succeed in developing shorter duration deepwater rice varieties or cold-tolerant boro varieties that allow early transplanting. In the double rice cropping pattern, the dry season rice yields are very high. This could add 8 million tones each year to rice production in Bangladesh and India. Breeding new rice varieties appropriate for the conditions is considered as the most favourable, effective, and efficient solution to increased rice production in adverse soils, as well as for adverse environmental conditions such as flooding, insects and diseases.

1.3 Why Boro Rice Cultivation is Gaining Popularity?

1.3.1 Adopted to Water Logged Lands

Boro rice cultivation is done mostly in waterlogged or low lying areas. In such lands traditional photosensitive deep water rice/flooding rice are grown in wet season with very low yield (0.5 to 1.0 t ha-1) and high risk. In dry season, any other crop could not be grown due to either water stagnation or excessive moisture. In these water logged 'chaurs' farmers are adopting boro rice crop as assured crop.

1.3.2 High Yield Potential

Boro rice cultivation is popular among the farmers remarkably due to its high yield potential, where a piece of land, giving 0.5 to 1.0 t ha-1 yield in wet season, produces 4-5 t ha-1 in boro season. Prabhat, Saroj, Richhari, Dhanlaxmi and Rajendra Bhagwati are adopted varieties for boro season. These varieties produce 1.5-2 times more yield in boro season than in wet season.

1.3.3 Risk Free Crop

Wet season crop in deepwater areas are vulnerable to drought and submergence but boro rice is free from these hazards.

1.3.4 Good Weather Condition

In boro season sowing period is October-November and the crop is harvested in April-may. During this season the crop enjoys more sunshine hours, lower night temperature, optimum temperature at ripening stage and environment partially free from insect pest and with a development irrigation system. All these result in higher yield potential of boro rice than Kharif. Though the crop duration is large, the corresponding yield is also high i.e. upto 10 t ha-1 reported in Bihar.

1.3.5 Animal Feed

In flood prone areas, flood is expected from July to September. During this period, it is very difficult to manage fodder for the farm animals. The straw obtained from boro rice in May/June proved a boon for the farmers of flood prone areas.

1.3.6 Profit

Produce of boro rice provides found economy to farmers. The income obtained from boro rice helps farmers to invest on inputs for next insuing crops.

1.3.7 Employment

Boro rice cultivation generates additional employment for farmers, labourers, and traders.

1.4 Limitation of Boro Rice Cultivation

1.4.1 Cold Injury

Boro rice grown in extensive areas of Assam, West Bengal and Bihar suffer from cold injury due to low ambient temperature during the early vegetative growth stage. The main constraint in boro rice area expansion is cold injury of rice at seedling stage. Providing cold tolerance genotypes through breeding strategies may give a quantum jump in boro rice endeavor. The term "cold tolerance" in case of rice refers to the capacity of plants to withstand temperature in the range of 15-20°C.

1.4.2 Conducive Environment For Boro Rice Cultivation

Boro rice is very location specific because of climatic conditions. The boro growing areas have some specific features different from the adjacent non-boro growing areas, in terms of some meteorological facts and available ground water. The specific characteristic in relation to boro cultivation are discussed below:

1.4.2.1 Rainfall

The area that receives more than 1000 mm annual rainfall proves to be suitable. Of the total rainfall, 10% should be distributed from January to April. It ensures high rate of establishment and early recovery of transplanted seedlings and maintains high humidity.

1.4.2.2 Temperature

Low temperature specially at seedling stage is a major constraint. But in boro rice growing areas, it does not go below 8°C and such temperature prevail for shorter periods than non-boro rice growing areas. The nursery remains surrounded by water. The temperature of nursery soil inundated by water remains more than atmospheric temperature. This situation is good for survival of seedling in boro rice growing areas, temperature starts to rise somewhat earlier, from mid-January to first week of February. This rise provides good opportunity for early transplanting and ultimately facilitate maturity before the onset of monsoon. Lower night temperature during the early stages of crop growth in winter facilitates the accumulation of photosynthesis, thereby increasing the carbon-nitrogen ratio. Favourable higher temperature at ripening period, resulting in higher production.

1.4.2.3 Humidity

High humidity (>70%) especially at flowering stage maintains pollen viability by nullifying the effect of high temperature (40-42°C). High humidity also minimizes water evaporation and provides conducive environment for growth.

1.4.2.4 Sunshine Hours

More sunshine hours favours to higher productivity of boro rice and minimize the pressure of biotic stress remarkably. The maximum yields are obtained in the dry season, when cloud cover is less and photosynthetic active radiation (PAR) is greater than during the wet season. Less insect/pest infestation on crop helps reduce damage.

1.4.2.5 Ground Water Level

High ground water table and sufficient availability of good quality water facilitate boro cultivation. In boro rice areas, ground water level is about 3 m below the surface. The bamboo boring system can easily establish to provide irrigation water to the crop. Higher water level also helps in maintaining the moisture regime in the field.

1.4.2.6 Soil

Clay and clay loam soil with high humus content is suitable for boro rice cultivation.

1.5 Suitable Varieties for Boro Rice and Their Characteristics Feature

Among the different agronomic practices, selection of a variety suiting to season is a pre-requisite for successful crop production. This is also true for boro rice. So in the selection of variety of boro rice one should must consider about cold tolerance at seedling stage as well as drought tolerance at flowering but cold tolerance is a major problem. Besides, variety should be short duration with high yielding ability. In the beginning farmers used traditional rice variety. They were tall in height, week stemmed, awned cultivars with poor in grain quality and poor yielder, as these were considered to have good tolerance to cold temperature at the seedling stage. Though the yield might not be high, they helped subsistence farmers. Chinsurah Rice Research Station in West Bengal conducted the experiment for the first time with their available germplasm and recommended Chinsurah Boro-1, Chinsurah Boro-2 boro season. Later on it was observed that photosensitive rice varieties also flower when grown in boro season. This opened a new avenue to develop better varieties than the traditional boro. Subsequently they were replaced by a number of varieties like Latisail, Patnai-23, Kalma-222, Jhinga sail and Raghu sail. In Bangladesh also four varieties namely, Habiganj II, Habiganj IV, Habiganj VI and Habiganj VIII were developed from local germplasm selections. Rice cultivars Topa boro, Jyoti Prasad, Mahsuri and Joymati are grown in Assam.

After introduction of dwarf high yielding short duration varieties, Pusa 2-21, Pusa 33 and Saket 4 became popular for boro season. But now a days medium duration varieties like Sita, Sujata IR 36 have also been adopted for this season of the varieties mentioned above were release boro rice. All these varieties have been adopted boro season only due to varying degree of resistance. It was great challenge before rice scientist to evolve a suitable rice variety with good varieties of cold tolerance and high yield potential exhaustive efforts of rice scientists, Rajendra Agricultural University, Pusa, Bihar has released boro rice variety GAUTAM in 1994. It is tall boro rice variety released in Bihar.

1.6 Characteristic Features of Some Popular Varieties of Boro Rice

1.6.1 Gautam

It is the first boro rice variety released by Rajendra Agricultural University, Pusa, Bihar in 1994. It has been developed through mutual breeding. It is a mutant of Rasi. It flowers in few days and mature in 195 days in boro season. It matures in 125-130 days only in Kharif season. It has excellent cold tolerance ability. In 0-9 temperature it has been scored 1-3 scale in different years. It average plant height is 70 cm. It has more nutrient of productive tillers per hill and high establishment ability of transplanted seed. Panicle length is about 25-28 cm with 180 numbers of grains per panicle. The harvest individual this variety is 0.48 to 0.52 with yield potential upto 10 t ha-1 in Bihar, and may be suitable for other states as well.

1.6.2 Pusa 2-21

It is a short duration dwarf high yielding variety released for irrigated upland condition for Kharif and Summer. But it has been adopted boro season also. In Kharif it matures in 100-105 days yielding plant height of 80-90 cm. It has moderate degree of cold tolerance. It takes 158 days to 165 days flowering and matures in 185 days in boro season. Its average plant height is 65 cm, panicle length 18-20 cm and number of grains per panicle is 118-120. The average yield potential in boro season is 5.4 t ha-1.

1.6.3 Prabhat

It is developed from cross IR 2033-251-1/IR 2612/ IR 36. It is semi dwarf variety matures in 90-95 days in Kharif season. Grain are long slender with yield potential of 5.0t/ha. It is also suitable for boro season. It has good degree of cold tolerance at seedling stage. In Kharif it is suitable for pre-flood and post flood situation. In boro it matures in 150 days.

1.6.4 Saroj

It is derived from Gautam/Type-3 cross combination. It is semi tally variety and it matures in 160 days in boro season and 110-115 days in kharif season. Its grain is long slender. It has good degree of clod tolerance at seedling stage and matures in 160 days in boro season. It has little aroma also.

1.6.5 Richharia

It is derived from IET 7564/Pusa 33 cross combination. It matures in 85 days in kharif and 165 days in boro season. It has tolerance to cold at vegetative stage and has synchronized flowering in boro season. It has yield potential of 5.0 t/ha and mature in 150 days. Its grain are long slender.

1.6.6 Dhanlaxmi

It derived from Saket 4/IR 36 cross combination. This variety matures in 100 days in kharif season. It has yield potential of 4.0t/ha.in kharif. It perform very well in boro season with yield potential of 6.5 t/ha and mature in 160 days. It has very good cold tolerance at seedling stage. It is resistance of BLB and zinc deficiency.

1.6.7 Rajendra Bhagwati

A newly developed culture RAU1397-18-3-7-9-4-2 has been identified from cross combination of IR36/Type-3. Its grain is long slender with good amount of aroma. It matures in 110-115 days and plants are semi tall. It is future variety of Boro rice having aroma. There is little information for identification of cold tolerant germplasm at seedling stage for boro season. Cold tolerance at flowering stage and seedling stage are independent and cold tolerance at seedling stage has been found to be governed by single dominant gene. Though variety with cold tolerance at seedling stage have been reported but their utilization in boro breeding programme have not been widely used. However survival of seedling is the most important in Bihar and UP due to low temperature, though there are other several constraints also. It was therefore, planed to screen a large number of germplasm of different maturity group for cold tolerance at seedling stage developed locally through hybridization and available through various sources.

The erratic rainfall pattern of the South-West monsoon comes in the way of timely field operation in rainy season. Also, the heavy rainfall during seedling in the field for a week or even more. In absence of any major environmental impediment during winter season, the desired field operations can be planned and executed at the most desired winter season, the desired field operations can be planned and executed at the most desired time. Moreover the absence of major diseases, insect-pests, etc. in this season, help in getting better profits from every additional unit of monetary inputs. As such the productivity of boro rice significantly exceeds the rainy season yield because of longer duration of crop growth enabling higher grain production. In this way, the crop can provide a major breakthrough for rapid increase in farm profitability. Its cultivation has become more popular after the development of high yielding hybrid and composite varieties with matching crop production technology. It has immense scope to be developed as an industrial crop.

Bihar, being one of the important rice growing states is grown in three seasons i.e. kharif, rabi (winter) and summer. During winter season an average grain yield of 4-6 tonnes/ha has often been obtained at farmers fields which is markedly higher than other season.

The boro rice with its high production potential has opened up a viable alternative during rabi and it is now a common practice to grow this crop in Bihar, Peninsular India and in north-eastern plains. Boro rice in Bihar has already substituted the cultivation of late sown wheat and has registered an impressive increase in production in recent years. Due to its potentiality to yield as high as 10-12 tonnes/ha and possibilities of further increase in substantial productivity with better management, the winter boro rice cultivation is increasing not only in Bihar but in other states also.

In Bihar, the average yield of boro rice is 2-3 tonnes/ ha in farmers fields, whereas, in experimental and demonstration plots the maximum of 4 and 6 tonnes/ ha has been recorded. Therefore, there is a wide gap in the yield of farmers fields compared to the experimental and demonstration plots. One of the important reasons for yield gap is partial/non-adoption of the production technology, our inability to involve large number of farmers in agricultural development programme and to equip them with adequate knowledge and skills necessary for efficient use of appropriate agricultural technology ad innovations. A large number of agricultural technologies are available which can be utilized by the farmers engaged in the crop production to their advantage. Included among these are high yielding varieties and production technology with respect to tillage, seeding rate, depth of sowing, seed treatment, placement of seed and fertilizers, hoeing and weeding, plant protection measures, cropping sequences and mixed/intercropping system etc. one of the ways to impart new knowledge and skills to the farmers is to have effective training programme for them. This is very important, because the complexity of enhancing food production is increasing day to day and the combination of factors limiting land productivity are continuing to get more complex. This requires a more holistic approach to meet the situation.

Training is assuming increasing importance as a crucial input for development in all walks of life obviously because humans constitute the most precious resource for development in any field, be it agriculture, industry, health or education. The world never had been so vast a store house of knowledge in different fields as it has today. The emerging knowledge and technology have become the strategic resource for accelerating the process of advancement. The greatest and the most urgent need of the day are to seize the opportunities they offer. This signifies the paramount importance of speedy and effective transfer of available knowledge and technology to the users. In fact, information has now become a critical input for rapid agricultural development.

The importance of training as an indispensable instrument for human resource development cannot be overemphasized. It aims at helping each individual to reach his maximum potential by way of increased knowledge, improved skills and changed attitudes enabling him to perform his job according to established standards. Training is the process by which an individual's efficiency and effectiveness can be maximized. It equips the individual with needed knowledge, attitudes and skills with respect to present or expected future roles and responsibilities enabling him to reach a desired level of performance. Farmers with high degree of awareness of the ground realities of agricultural systems and higher degree of motivation would only ensure qualitative change in the present level of production.

Prasad (1983) has remarked that our training programme as a matter of fact has been a piece-meal affair. The training designs do not include all related aspects of training, with the result of the effectiveness of training has not been appreciable. It is important to point out that agricultural training in India continues to be devoid of latest technologies in majority of the cases. Our training system has not gone beyond adhoc lecture of classroom approach. On the country, the industrial sector in India has been much more alive to changing training design and methodology. Therefore, it is high time that your training systems lay stress on quality training utilizing most of the effective approaches and methods.

Prasad and Mrutangey (1992) have stated that training is a critical input for the rapid transfer agricultural technology. Thus, the importance of training as an indispensable instrument for human resource development at any level cannot be ignored. In order to make any training meaningful and effective, the training needs have to be identified prior to the start of training programme. Choudhary (1991) has aptly remarked that every one speaks of the importance of training in agricultural but very little effort has gone into systematic identification of training needs. Verma and Verma (1985) have also rightly pointed out that all research attempts should concentrate on farmers' need and preferences.

The bright future for our agriculture would depend on knowledge intensive and science-based practice. Extension training is an integral part of agricultural production programme. Training builds up the professional competency of the farmers and extension personnel too. The farmers of lower socio-economic status with less education and knowledge about the various components of boro rice technology can also be well equipped through training programmes. The training needs of farmers and extension personnel need to determined in clear terms to provide proper direction and thrust to training efforts.

A large number of training programme for farmers were initiated right from the beginning of first five year plan to equip the farmers with the latest production technology and to bring positive behavioral changes in them. These programmes were: (1) special training of selected progressive farmers in the agricultural and allied fields; (2) training of progressive farmers in the intensive agricultural areas/blocks; (3) training of farmers in the non-intensive areas; (4) exchange of farmers within the country; (5) training of farm women in agricultural production, and (6) association of manufacturers, distributors, importers of agricultural inputs and exporters of agricultural products with the farmers.

1.7 Review of the Literature

Review of relevant literature is a fundamental and an essential research. Apart from determining the research need and delineation of the problem area, its main functions are to provide an insight into the methods and procedures, suggest operational definitions of main concepts and provide a strong basis for logical interpretation of findings. An attempt has been made in this chapter to gather some important findings of previous research work published pertaining to farmers training need which is the main subject of consideration in the present study. Some specific review of literature has been incorporated in the relevant chapters and sections of the thesis in order to make meaningful use of the same. The entire review of literature is classified in the four categories:-

(a) *The Training:* A Conceptual Analysis

(b) Training needs of farmers in new agricultural technology

(c) Opinion of farmers regarding training components

(d) Factor associated with training need.

1.7.1 Training: A Conceptual Analysis

Scholars had perceived the training from different perspectives. Hall (1962) defined training as process of aiding employees to gain effectiveness in their present or future work through the development of appropriate habits of thought, action, skill knowledge and attitude where as Lynton and Pareek (1967) described training as promises what is essential to modern technologies and economic system with flexibility in action through understanding and confidence.

Folley (1967) termed training as an overt process, a sequence of experiences, a series of opportunities to learn, in which the trainee is exposed in some more or less systematic way to certain materials or events. Whereas Dharma (1973) reported training as process to educate a person so as to be fitted, qualified, proficient in doing some jobs. For an extension workers, training includes the education which aims at bringing a desirable change in the behavior of the trainee of the learner.

Rao (1975) described farmer's training as an intensive learning activities for a group of selected farmers, assisted by competent trainers to understand and practice the skill required in adoption of new agricultural technology at a place where appropriate facilities exist and at time and duration considered suitable by the farmers. Similarly, Morrison (1976) described training need as existing any time when an actual condition differs from a desired condition in the human, or 'people', aspect of organization performance or more specifically, when a change in present human knowledge, skills or attitudes can bring about the desired performance.

Chaturvedi (1979) reported that broadly, training aims to maintain and increase the employee's effectiveness in his present job, prepare him for promotion by stimulating his potentials, and develop his skills and knowledge for greater organizational effectiveness. In the same way, Lynton and Pareek (1990) described four assumptions underlying new concept of training:

1.7.1.1 Motivation and Skills Lead to Action. Skills are acquired through practice.

1.7.1.2 Learning is complex function of the motivation and capacity of the individual participants, the norms of the training group, the training methods and the behavior of the trainers and the general climate of the institution. The participant's motivation is influenced by the climate of his work organization.

1.7.1.3 Improvement on the job is a complex function of individual learning, the norms of the working group, and the general climate of the organization. Individual learning, unused, leads to frustration.

1.7.1.4 Training is the responsibility of three partners: the participant's organization, the participant and the training institution. It has a preparatory, pre-training and a subsequent post training phase. All are of key importance to the success of training.

1.8 Training Needs of Farmers in New Agriculture Technology

Extension Education Institute, Rajendra Nagar (1971) analyzed the training needs of farmers in relation to new agricultural technology and reported training in Plant protection as the foremost need of farmers followed by use of fertilizers and new seed varieties.

Satynarayan and Bhaskaran (1971) reported that the knowledge and skill of farmers needs to be improved were: determination of fertilizer doses and their application, soil sampling, use of plant protection chemicals and appliances, identification of pests and diseases and soil control. The study also indicated that the farmers needed specific training in all the package of practices of high yielding varieties except land preparation, drainage and harvesting operations. Jha (1974) revealed that the medium farmers perceived training in plant protection measures as 'most needed', while small farmers perceived training in plant protection measures and high yielding varieties of paddy as 'most needed'. Haque (1975) analyzed the training needs of farmers in relation to high yielding varieties of paddy and reported that training in plant protection measures was perceived as 'needed'—and high yielding varieties, fertilizers, suitable improved implements, and nursery raising perceived to be as, 'somewhat needed'.

Pandey and Singh (1976) revealed that farmer of irrigated and non-irrigated areas perceived training need in high yielding varieties of wheat including plant protection and fertilizers application 'most needed'. Training in seed treatment and land preparation was perceived as moderately needed. Singh and Singh (1981) analyzed the training needs of tribal farmers. Some of the salient findings were as follows:—

1.8.1 The main areas in order to their training needs were agriculture, animal husbandry, poultry, bee keeping, storage of food grains, marketing of produce and gobar gas plant.

1.8.2 The sub-areas of training in the field of agriculture in which training was 'most needed' were name and characteristics of recommended high yielding varieties, construction of channels/drains and seed treatments.

1.8.3 As regards the sub-areas of training in the field of storage of food-grains, knowledge of improved methods of storage and type of storage bins were ranked first and second repetitively in respected of their training need.

Anandan and Vasanthakumar (1999) while analyzing the training needs of groundnut growers observed that the respondent needed training in the major subject matter areas of fertilizer management, pant protection, seeds and sowing and weed management. The irrigated groundnut also expressed that they did not need training in cropping system.

1.9 Opinion of Farmers Regarding Training Components

1.9.1 Time of Training

Srivastava (1963) found that most suitable month for organizing training camp were January and August. Whereas Sanders (1967) stressed that training for kharif should be organized in the month of April-May because it is slack season for farmers of Gujrat. Similarly rabi camp should be organized in the month of September. Similarly, Sidhu and Patil (1968) reported that to make the training programme effective, the training camp for kharif should be organized in the month of April-May and rabi in the month of September-October with the duration of camp being to be of two to three days.

Sohal and Yanakani (1970) analyzed the training programme for young farmers organized by the Punjab Agricultural Univeristy and reported that it should be organized during the month of January, February and March and the duration should be of three months the course should be held at the Punjab Agricultural University. Where as Jha (1974) analyzed the preference of the farmers for the various components of farmers training programme and reported that the training for high yielding variety of paddy organized before the onset of sowing time, most preferably in the month May and June at the village level was likely to be very popular among the paddy growing farmers.

Sinha (1998) opined that after combining the choice of farm-women in respect of different components on top priority it emerged that marginal farm-women and small farm-women had similar choice for organizing training in village during the slack season with demonstration method for ½ day duration. He further reported that medium farm-women had shown their desire for organizing training in village before the onset of crop season adopting demonstration method for 3 days' duration.

1.9.2 Duration of Training

Barooh (1965) suggested that the duration of the training should be of one day to three days. Whereas

Monitoring and Evaluation Unit, Department of Agriculture, Bihar (1984-85) analyzed the duration of training programme and it was reported to vary from 4-6 hours but in some divisions, the duration of training programme was found inadequate.

Kumar (1985) reported that majority of farmers preferred to organize the training two times in a year while approximately one-third favoured once in a year. One day training was preferred mot by the farmers. Haque (1992) reported that duration of training should be of 2 days and the programme should be organized from 12 noon to 4 pm. Choudhary (1999) while analyzing the opinion of farmer regarding the duration of training reported that the marginal farmers had preference for organizing training during the slack season for two days duration at a frequency of once in a month. Whereas, small and medium farmers had shown their desire for organizing training during the crop season for 3 days duration and frequency of training should be once in a crop season.

1.9.3 Size of Trainees

Geues (1964) reported that convenient size of training group in a village was about 20 farmers. Whereas Workshop on farmers training and education (1967) recommended that optimum size of training group at the village level should be 30-40 members. Similarly, Rao (1969) studied farmers training programme in intensive agricultural district programme and intensive agricultural area programme districts in relation to high yielding varieties programme and reported that the usual size of farmers training group was 50-100 and optimum size preferred was 30-50. Sohal and Yanakani (1970) analyzed the training programme for young farmers organized by the Punjab Agricultural University and reported that a group comprising 25 to 30 trainers should be an appropriate size.

1.9.4 Venue of Training

Sanders (1967) suggested that for creating better conviction in the mind of the farmers for innovations and also not to disturb their village programme it should always be desirable that raining must be organized in the villages. Sidhu and Patel (1968) observed in their study of opinion of trained farmers and concerned officials on the organization of farmers training camp and at village level it had gone a long way inspiring the farmers to adopt improved farm practices. Sidhu and Patil (1968) reported that farmers' training camp at village level had gone to long way in inspiring the farmers to adopt improved farm practices. Jha (1974) and Pandey and Singh (1976) from Bihar revealed that small farmers preferred training programme to be organized in their own villages. Similarly view was echoed by Haque (1992) while analyzing the opinion of farmers regarding components of training programme suggested that the venue of the training programme should be at farmers fields and villages.

1.9.5 Trainers Category

Haque (1992) while analyzing the opinion of farmers regarding training components reported that majority of the respondents opined that training should be imparted by rice specialists of Rajendra Agricultural, Bihar, Pusa and Sr. Officers of Department of Agriculture.

1.9.6 Training Methods

Srivastava (1963) found that the method of training should include demonstration, field trip, group discussion and practical. Barooh (1965) also suggested similar views. Workshop on farmers training and education (1967) recommended the method and result demonstration, group discussion and field trip to be used for training programme. The content of training programme should include technical content, economic and risk in using the package of practices, credit and supply. Sidhu and Patil (1968) reported that more emphasis should be laid on practicals instead of lectures and opportunity to practices should be provided to the farmers. The field trip should be undertaken to the research station, seed farms, cultivators field and demonstration plot. Exhibition should be organized and literature be supplied to the farmers.

Mangat and Hansra (1988) observed that for gain in knowledge of specific facts like dates, events, persons, places, magnitude of the phenomenon and other approximate or relative information, demonstration and discussion was the most effective method combination followed by film and discussion, slides tape and discussion and lecture and discussion.

1.9.7 Follow-Up Activities

Singh (1967) felt that the general approach to training should be problem oriented. The farmers should be shown, as far as possible ways to solve their problem. He further pointed out that well-planned follow-up programme for the training farmers would help in achieving the desired goals to a considerable extent. Similarly, Haque (1992) reported that as follow-up activities, majority of the respondents suggested that demonstration should be conducted in farmer's fields and extension workers should contact the trained farmers.

1.10 Training Needs in Rice Technology

1.10.1 The variable age was found to be positively and significantly associated with rank variability in training need in all the categories of farm women.

1.10.2 Variables like value orientation, economic motivation, size of holdings, and age were found to have positive and significant association with training need in small category of farm women.

1.10.3 In medium category of farm women variables like value orientation, economic motivations and size of holding were found to have negative and significant association with training need.

Choudhary (1999) in his study on training need of farmers in relation to high yielding rapeseed and mustard in adopted villages reported that perception of farmers regarding association of training need with independent variables. Some of the salient finds were:

1.10.4 Education, knowledge of the farmers, economic motivation and cropping intensity were positively and significantly associated with the training need of all categories of farmers.

1.10.5 In case of marginal farmers, size of the holding was found positively and significantly associated with the training need of farmers.

1.10.6 The variable namely, irrigation potential, contact with extension agencies and mass-media exposure were also positively and significantly associated with the training need of all farmers.

Even after sixty year of Independence the India is still agricultural country. 78 per cent of population lives in villages. About sixty five per cent of Indian population depends on agriculture as profession for their livelihood. Among the farming population a large number of farmers in our country come in the category of small and marginal farmers and operate on approximately 38 per cent agricultural land. Any policy decision or strategy thus, should certainly attract the consideration for long term interest of these farmers. The option which may help Indian farming community is to equip our extension personnel and different categories of farmers with latest knowledge and skills to meet new requirements, and take advantage of synergy through the efforts of farmers and efficient extension machinery. Also because of new developments in international trade, globalization and shifts in agricultural cropping patterns from subsistence to commercial agriculture, the situation demands new extension approach for benefiting the farming community, specially the small and marginal farmers. Looking to the emerging needs for never knowledge and skills of farmers in the new era, the extension system has to play a vital role and equip itself in developing farmers as agripreneurs. Then only we can meet the future challenges in farming sector.

Therefore, it is urgently needed to arm the scientists and extension workers with the power of knowledge about training needs of different categories of farmers and extension workers. There is need to develop sound training strategy to bridge the gap between the boro rice production technology available at the research station and farmers' field. This will translate the concept of 'precision farming', which is backed by knowledge intensive farming practice for sustained growth.

Thus, realizing the importance of the problem, the present study was undertaken for the assessment of training needs and designing strategy for increasing boro rice productivity with the following objectives:

a) To identify agro-economic, socio-psychological and communicational characteristics of boro rice growing farmers.

b) To analyze the extent of technological gap in adoption of scientific boro rice production technology.

c) To assess the training needs in respect of scientific boro rice production as perceived by farmers and extension personnel.

d) To find out the association of the training needs with certain selected agro-economic, socio-psychological and communicational variables.

e) To seek the opinion of farmers in respect of different components of training programme.

f) To develop training strategy for different categories of farmers for increasing boro rice production.

1.11 Importance and Scope of the Study

Looking into the development of Indian agriculture, and changes taking place in domestic as well as international environment, the strategies and methodologies for agricultural extension need to be modified. Even the definition and scope of agricultural extension needs to be redefined. Subjects for attention in the new millennium are the issues related with land, its productivity and optimum utilization, fertilizer and water management for economic use, marketing and processing of produce and group driven management. Economic prosperity of farmers with focus on 'rupees generating rupee' instead of 'seed to seed' has to take front seat in driving agricultural extension in 21st century. Farmers are now not only the producers of crops commodities but they are also to play a vital role as agri-preneurs in coming years for which they will certainly require support from the agri-extension system. A technology driven, science based and industry linked production base hastening the much needed breakthrough for the Indian boro rice economy is the need of the hour.

Although certain programmes for promotion of boro rice like special food grain production programme (SFPP), distribution of mini-kits to the farmers, conducting front-line demonstrations under technology mission etc., have been initiated/implemented by the Government, concerted efforts are required to make the crop more productive and remunerative. One important way out in this direction would be to develop farmers knowledge base and skill with high degree of awareness of the ground realities and technological advancements.

The present study will generate useful information on the training need of farmers for increasing boro rice productivity. Farmers have to be equipped with the latest technologies so that they can take full advantage of technological advances. The only way we can make the best use of the actual and potential capacities of farmers is to provide them opportunities for self development through training. Preparatory essential to any training programme should be to determine in clear terms, the training need of farmers. This is necessary to provide desired direction and push to training efforts. There stands a good scope of increasing production by way of improving the knowledge as well as skills of the farmers in using the modern inputs efficiently. Keeping this in view, the present study was planned to generate useful and meaningful information on various aspects of training needs in terms of different categories of farmers. It is expected that the finding of the study will help the government as well as the voluntary and non-voluntary organizations engaged in training of farmers to organize training programmes on the basis of felt needs and interests. The knowledge of training needs of different categories of farmers will help the institutions/ organizations to formulate suitable training programmes and course curriculum for different categories of farmers to meet their differential training needs. The findings of the study will help in assessing the existing technological gaps in boro rice cultivation and also in designing the strategies fulfill the gap with technological know how. The findings will also help to know the view of the farmers about different components of training programme i.e., venue, time, duration and training methods like lecture, question-answer, tour, field visit, exhibition, publication etc.

In order to have comparative view of different categories of farmers in respect of training needs in crop production technology, the present study will have definite bearing on developing training strategies for different operations in boro rice production.

1.12 Scope of the Study

The present study had the limitations of time and resources. However, considerable care and thought were exercised in making the study as objective and systematic and purposeful.

The study involved collection of data from boro rice growing for farmers augmenting boro rice productivity and information relating to a number of variables selected for the study. The correctness of the responses, which were based on recall, in spite of best efforts of the investigator, might have left margin for error to creep in.

It may, however, be pointed out that the findings of the study should not be generalized beyond the limits of the area under investigation and other areas having similar agro-climatic and socio-cultural conditions. In spite of the limitations, it is hoped that the findings of the study would be useful to the Boro rice scientists, planner's, extension workers and various organizations engaged in training programmes.

1.13 Layout of the Study

The thesis has been clubbed together into five chapters. The first chapter as usual is introduction. Apart from introduction it deals with the review of literature related to the present study. In the second chapter, the methodology, deals with the methodology used including the measurement of different variables has been described. The third chapter deals with the research setting of the study. The analysis and interpretation of data and discussion of findings constitute the content of the fourth chapter. The summary and conclusion have been presented in the fifth chapter.

2

Methodology

This chapter deals with the methods, procedures and techniques used in the present study for collecting and analyzing data with respect to the objectives formulated. This has been divided into four parts/sections.

(A) Sampling techniques used.

(B) Empirical measures of dependent and independent variables

(C) Techniques of field data collection and

(D) Statistical tools used for analysis and presentation of findings

A. SAMPLING TECHNIQUE

2.1 Selection of District

Madhubani district was purposively selected for the purpose of study in view of its importance in terms of area and total production of Boro ries. This district is predominantly an agricultural district and represents basically a rural agricultural economy. This district holds an important place in Bihar, since agricultural development in this district has taken a long strides in recent years and soil and climate of this district are also favourable for increasing production.

2.2 Selection of Block

There are twenty one blocks in Madhubani district. Names of the blocks were noted from records maintained at the district headquarter. Out of twenty one blocks two blocks i.e. Madhepur and Lukhnaur were purposively selected for this study.

2.3 Selection of Respondents

The farmers in each of the two selected block were stratified into four categories on the basis of their size of land holding viz. marginal (Below one hectare), small (one to two hectare), medium (two to four hectares) and large (above four hectares). The criteria adopted by the Expert Committee on Assessment and Evaluation (1980) of the Ministry of Food and Agriculture were adopted. As the total number of large farmers in the selected villages were very few, the medium and large farmers were grouped together and categorized as medium farmers in the present study.

A list of farm families engaged in boro rice cultivation was prepared on the basis of their cultivated land with the help of block officials and local leaders. 50 respondents from each village were randomly selected. 200 farm families were selected for the study, but their selection was proportionate to the size of farm families under each category. Thus the total number of respondents selected was 200 under three categories of marginal, small and medium farmers respectively from the selected sample village. The number of selected respondents in each village and block under different categories have been given in Table 2.1.

Table 2.1: Distribution of Respondent Blocks

S.No.	*Block selected*	*No. of households*	*No. of farmers selected for study*	*Category of farmers*		
				Marginal	*Small*	*Medium*
A.	Madhepur	210	100	60	27	13
B.	Lakhnaur	278	100	59	27	14
	Total	488	200	119	54	27

2.4 Empirical Measures of Dependent and Independent Variables

On the basis of the content analysis of relevant past studies and available other literature the variable frequently reported in affecting training need were collected. This was supplemented by informal discussion with scientist. Further the variable so identified was discussed at length in the advisory committee and again they were presented before scientist/faculty members in the seminar specially held to finalise the research project. After in-depth discussion the selected independent variables were categories in five agro-economic, seven socio-psychological and two communicational variables. The variables and their measurement are given below in Table 2.2.

Table 2.2: Variables Selected for Study and their Measurement

Sl.No.	Variables	Measures
A.	DEPENDENT VARIABLE	
1	Training Need	Average Choice Score (ACS) Formula, suggested by (Singh, 1980)
B.	INDEPENDENT VARIABLES	
	Agro-Economic	
1	Farm size	Structured schedule
2	Irrigation index	Schedule developed in this study
3	Cropping intensity	Structured schedule
4	Farm mechanization	Farm machanisation index (Chaudhary, 1983)
5	Technological gap	Structured schedule
	Socio-psychological	
1	Age of farmer	Structured schedule
2	Education of farmer	Socio-economic status scale rural (Pareek and Trivedi, 1964)
3	Family size	Structured schedule
4	Knowledge level	Test developed in this study
5	Risk orientation	Risk preference scale (Supe, 1969)
6	Economic motivation	Economic motivation scale (Moulik, 1965)
7	Market orientation	Management orientation scale (Samanta, 1977)
	Communicational	
1	Utilisation of mass media	Schedule developed in this study
2	Utilisation of personal cosmopolite	Schedule developed in this study

2.5 Dependent Variable

2.5.1 Training Need

The study aimed at studying the training needs of the farmers in relation to boro rice production and the factors influencing this training need in the context of agricultural and rural development are a little different as they are mostly related to preparation of people for new facts of technology or, development and adoptions of innovations or new practices in relation to agricultural technologies and development efforts. In this study training need has been conceived as the gap between what is and what ought to be in actual performance of the task relating to boro rice cultivation. In the present study training need of the farmers of different categories engaged in boro rice cultivation was studied against the most significant main areas and their sub-areas of training.

In order to determine the main areas of training in relation to boro rice production both primary and secondary sources of information were used. The primary sources of information included field observation and the consultations with the field extension workers as well as the concerned scientists of the university and farmers, whereas, publications report books and journals served as secondary source of information. Before selecting the ten main areas of training in relation to bori rice production technology a list of forty six main areas were prepared with the help of literature related to the concerned crops and the farmers training need.

Forty-six sub-areas of training in relation to boro rice were taken which is mentioned below in Table 2.3.

Table 2.3: Main Areas and Number of Sub-Areas Under Each Main Areas of Training

Sl.No.	*Main areas of training*	*No. of sub-areas of training*
1	Field preparation	3
2	Use of HYVs	4
3	Seed treatment	3
4	Improved cultural practices	8
5	Integrated nutrient management (INM)	5
6	Weed management	6
7	Cropping system	3
8	Integrated pest management (IPM)	5
9	Water management	5
10	Post-harvest technology	4
	Total	46

The training need of farmers in the main areas and sub-areas of training in relation to winter boro rice as perceived by them were measured by using a three point rating scale i.e. first choice, second choice and third choice and it was quantified by assigning score 3, 2 and 1 respectively, and the respondent were asked to respond in specific items in a three point continuum and frequency of response multiplied with corresponding score and added them they were divided with the number of continuum which give average choice score (ACS).

The perception of training need of extension personnel was worked out by asking them to rank the training area in boro rice technology for all the three categories of farmers in the line what described in the preceding part of the section and there perception were compared with the perception of different categories of respondents with the help of rank order correlation.

2.6 Independent Variables

2.6.1 ***Agro-Economic Variables:*** Like Farm size, irrigation index, cropping intensity and farm mechanization was used.

2.6.2 ***Technological Gap:*** It has been operationalised as the proportion of recommended technology not adopted by the farmers. This can be determined only after proper adoption is computed. Technological gap in the cultivation high yielding varieties of boro ries has been conceived as the difference between the recommended package of practices and the extent of adoption of recommended package of practices. The technological gap was calculated for each of the item of the package of practices. The extent of adoption was conceived as the degree to which a particular practice for boro rice cultivation was adopted by a respondent.

2.6.3 ***Collection of Items:*** The content of a technological gap is composed of question called items. A large number of items with respect to boro rice production were collected in consultation with boro rice scientists research workers of agronomy department and field extension workers.

2.6.4 ***Procedure of Scoring:*** For quantifying the technological gap of the respondents, one score was assigned to right answer and zero score for wrong answer or no answer in respect of each item of every question of the technological gap. The mean scores obtained and technological gaps were worked out. Based on the percentage gapes, the ranks were assigned to each practice. A list of selected main practices are presented below:

a) Improved seed and their treatment

b) Sowing methods

c) Fertilizer application

d) Irrigation

e) Pest control

f) Weed control

g) Post harvest technology

The technological gap was computed by suing the following formula:

Technology gap =

Extent of recommendation – Extent of adoption

x 100

Extent of recommendation

2.6.5 ***Age of the Farmer:*** In the present study the number of years rounded to the nearest whole number the respondent lived since birth at the time of interview was taken as a measure of age of the farmer.

2.6.6 ***Education of the Farmer:*** It refers to the respondents academic attainment through formal schooling. To quantify the educational status of the respondents, the scoring system followed by Pareek and Trivedi (1964) in their socio-economic status scale – rural was used in this study. The scoring was as follows:

Level of Education	*Score*
Illiterate	0
Can read only	1

Can read and write	2
Primary	3
Middle school	4
High school	5
Graduate	6

2.6.7 ***Family Size:*** It refers to the number of members in the individual farmer's family. To quantify the family size, actual number of members in the family was taken into consideration.

2.6.8 ***Measurement of Knowledge:*** The Webester's dictionary defined 'knowledge' as acquaintance with facts, range of information, awareness or understanding, the body of facts, principles etc. accumulated by mankind as far as one knows within the range of one's information.

Knowledge as defined in this study include 'those behaviours and test situations which emphasized the remembering either by recognition of recall of ideas, material of phenomenon' (Bloom and other, 1956).

Knowledge is the totality of understood information possessed by a person. Rogers and Shoemaker (1971) identified three types of knowledge — awareness knowledge, how to knowledge and principle knowledge. A standardized knowledge test based ion the procedures described by Lindquist (1951) was developed to measure the knowledge of farmers. The variable indicated the extent of knowledge the respondent possessed at the time of interview as evident from his responses to a set of questions scientifically prepared for this purpose. The steps followed by developing the knowledge test for this study are discussed here.

2.6.9 ***Item Collection:*** The content of knowledge test is composed of questions called item. A large number of items with respect to package of practices of high-yielding varieties of paddy were collected in consultation with specialists, trainers, extension workers and farmers of the area of research. The items were collected in relation to HYVs, land preparation and sowing, seed treatment, fertilizer management, water management, plant protection, weed control, intercropping with boro rice, use of farm machinery, harvesting and storage and transplanting of boro rice were taken into consideration.

2.6.10 ***Item Analysis:*** Item analysis of a test yields two kinds of information i.e., an index of item difficulty and an index of item validity. The selected 91 statements were administered to 20 farmers randomly selected who were different from the main sample of farmers selected for this study. Responses were dichotomized as 'correct' and 'incorrect' or 'don't know'. If a farmer gave correct answer to an item he was given a score of one and those who gave incorrect answers or told don't know got zero score for the item. The total scores for each farmers was thus calculated. Afterwards the scores were arranged in descending order of scores from highest to lowest. As suggested by Harper JR et al. (1964) twenty seven per cent of farmers with highest score and 27 per cent with lowest score were selected for calculation of item difficulty and validity.

2.6.11 ***Final Selection of Items:*** Two criteria, namely, item difficulty index and item discrimination index were considered for the selection of items in the final format of the knowledge test.

2.6.12 *Method of Scoring:* The respondent were asked to answer the items in dichotomized from like 'correct-incorrect', 'yes-no' etc. One score was given for each correct reply. The summation of scores for correct replies overall the items of a particular respondent indicated his level of knowledge about boro rice production technology. The range of scores was from 0 to 71. The knowledge test had eleven components and the summation of scores for correct replies overall the items of a particular component indicated the level of knowledge of the respondent about that component. The maximum scores attainable for the components were knowledge about HYVs (8), knowledge about land preparation and sowing (7), knowledge about seed treatment (4), knowledge about fertilizer management (9), knowledge about water management (6), knowledge about plant protection (12), knowledge about weed control (4), knowledge about intercropping with boro rice (3), knowledge about farm machinery (6), knowledge about harvesting and storage (4) and knowledge about transplanting of boro rice (8).

2.6.13 *Measurement and Analysis of Training Component:* The opinion of the farmer regarding various components of training namely venue, time, duration, number of trainees, method of training, combination of training, trainers categories and follow up activities for organizing effective training programme were analysed. A list of specific items under each components were prepared on the basis of review of available literature and informal discussion with farmers of the areas of investigation. A four point rating scale was used to seek the opinion iof the farmer and regarding specific items of the various components of training programme. The four points rating scale with their score given in parenthesis were most preferred (4), moderately preferred (3), less preferred (2) and net preferred (1).

2.7 Techniques of Field Data Collection

Before the construction of data collection device a pilot study was undertaken in the villages of two blocks i.e. Madhepur and Lakhnaur of Madhubani district.

2.7.1 Construction of the Schedule

The draft schedule for collection of data incorporating the tools and techniques of measurement of different variables was pre-tested twice each time with boro rice growing farmers. In pre-testing care was taken not to include farmers who were selected as sample for final interview.

One the basis of experience in pre-testing, appropriate changes in the construction of questions and their sequence were made. The schedule was then finalized and duplicated. The final from the interview schedule is given in Appendix-II.

2.7.2 Field Data Collection

The data were collected with the help of schedule constructed for the study. In each block before starting the interview a few days were devoted to the establishment of support with the respondents.

The schedule was administered in local language (Maithali) and replies were recorded in English on the schedule. The interview was done by the researcher himself.

2.8 Statistical Methods Used

The collected data's were analyzed and interpreted by using standard statistical methods.

3

Research Settings

Bihar after division in 1999-2000 is dominantly an agriculture-based state, where lands are mostly plain, soils are deep and climate is monsoonic with highest water and labour resources. These five factors coupled with diversified type of land use choice in any given land unit, are favorable to promote agricultural development not only to compete with the food requirements of the rapid growing population, but also rural economic growth in Bihar with system based planning. Other than crops, the land is potentially rich for livestock production and fishery. However, agro-forestry deserves promotion in different facets in Bihar.

3.1 Location and Extent

The state of Bihar is located between 24^0 17′ and 27^0 31′ N. latitude and between 83^0 19′ and 88^0 17′ E longitudes covering an area of 9.38 mill. hectares which accounts for nearly 2.86 percent of the total geographical area of the country. Its northern boundary forms the international boundary with Nepal. It is bounded by U.P. and M.P.

in the west, Jharkhand in the south and West Bengal in the east. The state has 38 administrative districts with Patna as the capital. The average length East to West is 483 km and North to South is 385 km, 38 districts is represented by 40 parliamentary constituencies and 243 legislative seats.

3.2 Land Utilization

Land use or land utilization type can broadly be grouped into two categories viz. (a) non-arable and (b) arable. Non-arable land is classified as forests, barren and unculturable land, land put to non-agricultural uses, permanent pasture and grazing land, miscellaneous trees and groves, current fallows and other fallows. Arable land includes all cultivated area put under various crops. District wise land utilization pattern and district wise classification of crops as practiced in the sate are also shown in Table 3.1.

3.3 Climate

The climate is subhumid subtropical monsoon type. The mean annual rainfall varies from 200 mm near the northern and northern corner of sate of a minimum of 100 mm on the western border along with Ganga axis. The mean annual temperature varies from 24° to 26° C with mean summer temperature ranging between 28 and 33° C and mean winter temperatures between 17° and 19° C. The tables 3.1(a) & 3.1(b) gives the details of meteorological and temperature data of boro rice glowing areas of Bihar.

Table 3.1: District wise and Zone wise Classification of Land in Bihar

(In 000'Hectare)

Sl. No.	*District Name*	*Total geographica area by village paper*	*Forest*		*Barren and unculturable land*		*Land put to non-agricultural uses*		*Permanent pasture and other grazing land*	
	Zone-I		*1997-78*	*1998-99*	*1997-78*	*1998-99*	*1997-78*	*1998-99*	*1997-78*	*1998-99*
1.	West Champaran	484	92	92	3	3	89	93	1	1
2.	East Champaran	432	0	0	8	8	63	63	1	1
3.	Sheohar	43	0	0	0	0	12	12	0	0
4.	Sitamarhi	222	0	0	2	2	52	57	1	
5.	Madhubani	353	0	0	2	2	88	88	1	1
6.	Darbhanga	254	0	0	1	1	61	58	0	0
7.	Muzaffarpur	315	0	0	5	5	62	62	0	0
8.	Gopalganj	204	0	0	6	6	32	32	0	0
9.	Siwan	224	0	0	9	9	30	31	0	0
10.	Saran	265	0	0	18	18	32	33	0	0
11.	Vaishali	201	0	0	24	24	37	36	0	0
12.	Samastipur	262	0	0	4	4	63	63	0	0
	Zone Total	3259	92	92	82	82	621	628	4	4

	Zone - II									
13.	Begusarai	188	0	0	18	18	44	44	0	0
14.	Supaul	239	0	0	20	20	48	31	0	0
15.	Araria	272	1	1	5	5	53	51	0	0
16.	Kishanganj	189	0	0	11	11	35	36	0	0
17.	Purnea	314	0	0	12	12	45	45	0	0
18.	Katihar	291	2	2	22	22	58	59	0	0
19.	Madhepura	180	0	0	4	4	41	31	1	1
20.	Saharsa	165	0	0	11	11	32	31	1	1
21.	Khagaria	149	0	0	14	14	30	30	0	0
	Zone Total	1987	3	3	117	117	386	358	2	2
	Zone-IIIA									
22.	Bhagalpur	254	0	0	23	23	62	67	1	1
23.	Banka	306	46	46	43	43	43	42	2	2
24.	Munger	140	29	29	12	12	29	31	0	0
25.	Lakhisarai	129	13	13	7	7	13	13	0	0
26.	Sheikhpura	61	0	0	1	1	9	10	0	0
27.	Jamui	305	93	93	29	29	38	38	2	2
	Zone Total	1195	181	181	115	115	194	201	5	5

	Zone-IIIB									
28.	Nalanda	234	5	5	1	1	41	41	0	0
29.	Patna	317	0	0	13	13	75	75	0	0
30.	Bhojpur	237	0	0	7	7	32	33	0	0
31.	Buxar	167	0	0	2	2	16	16	0	0
32.	Bhabhua	342	113	113	19	19	33	33	0	0
33.	Rohtas	391	67	67	17	17	47	47	0	0
34.	Jehanabad/Arwal	157	1	1	5	5	24	24	0	0
35.	Aurangabad	330	14	14	16	16	55	55	1	1
36.	Gaya	494	78	78	28	28	67	72	3	2
37.	Nawada	249	64	64	11	11	34	35	1	1
	Zone Total	2918	342	342	119	119	424	431	5	4
	State Total	9360	616	616	437	437	1616	1635	16	15

Source: 25 years perspective plan of land use planning of Bihar

Table 3.1(a): Meteorological Data of Boro Growing Areas of Bihar (Based on Average of 20 Years Data from 1971 to 1990)

Months	*Rainfall (mm)*	*Temperature (oC)*		*Average relative humidity*	*Average sunshine hours*
January	11.7	8.2	22.8	92	7.4
February	17.2	9.6	24.8	90	8.6
March	14.2	13.8	30.3	82	8.6
April	52.5	18.7	33.3	79	8.6
May	116.9	22.1	33.0	85	8.6
June	256.1	24.2	33.9	90	5.2
July	459.0	24.5	31.9	92	5.2
August	288.7	24.9	31.6	90	5.6
September	138.8	23.8	31.4	92	6.4
October	67.2	20.9	28.3	92	8.2
November	8.8	14.6	27.1	91	8.4
December	7.7	9.7	23.9	92	7.5
Total	1438.8				

Table 3.1(b): Temperature Available at Different Growth Stage of Boro Rice in Boro Growing Areas of Bihar

Growth stages of boro rice	*Temperature*					
	Standard fixed by IRRI			*Temperature available in boro growing area of Bihar*		
	Min.	*Max.*	*Av.*	*Min.*	*Max.*	*Av.*
Germination	10	45	20-35	14.0	31	20-28
Vegetative growth	12-13	35	25-30	14.1	20	12-20
Tillering	9	33	25-31	12.0	27	15-25

Ear development	15-20	38	--	16.0	31	17-32
Flowering	22	35	30-33	22.0	36	21-35
Grain maturity	12-18	30	20-25	25.0	40	24-38

Figure 3.1

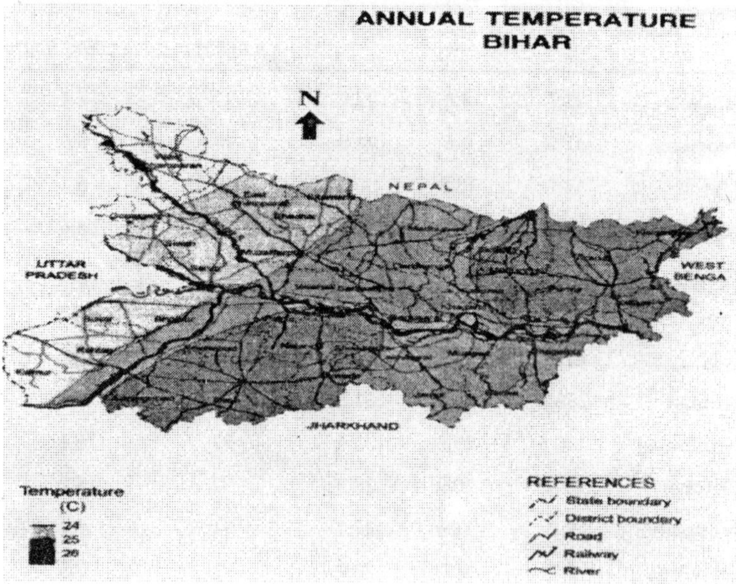

Figure 3.2

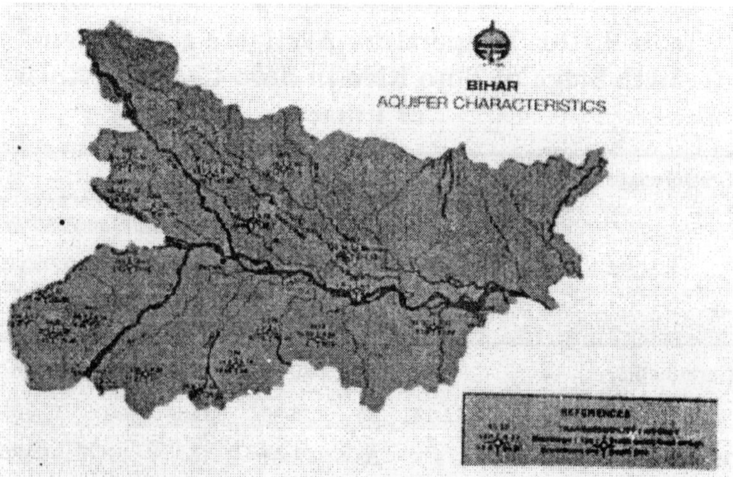

Figure 3.3

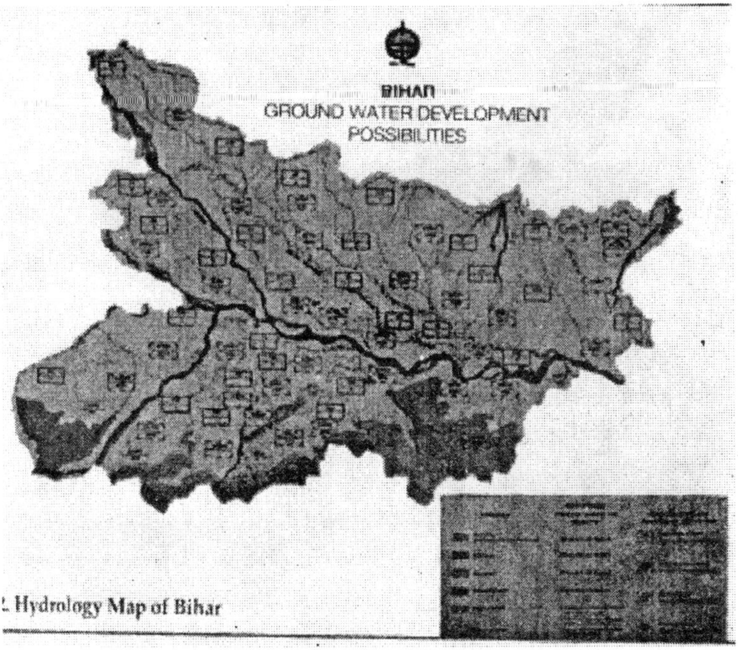

Figure 3.4

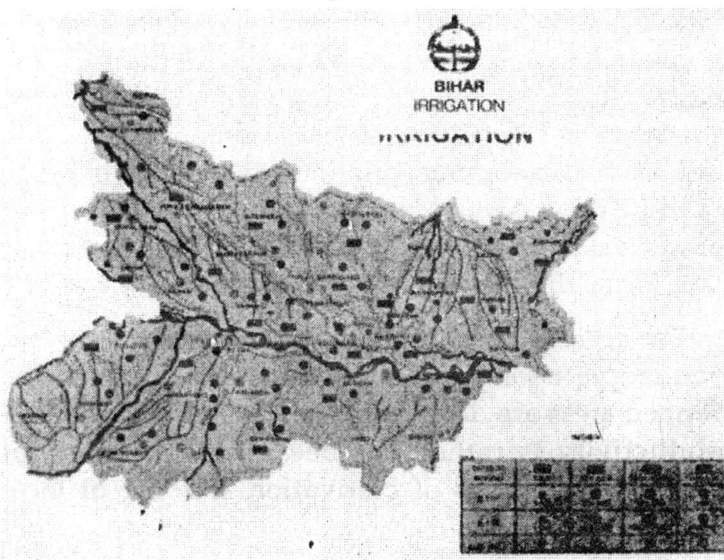

Figure 3.5

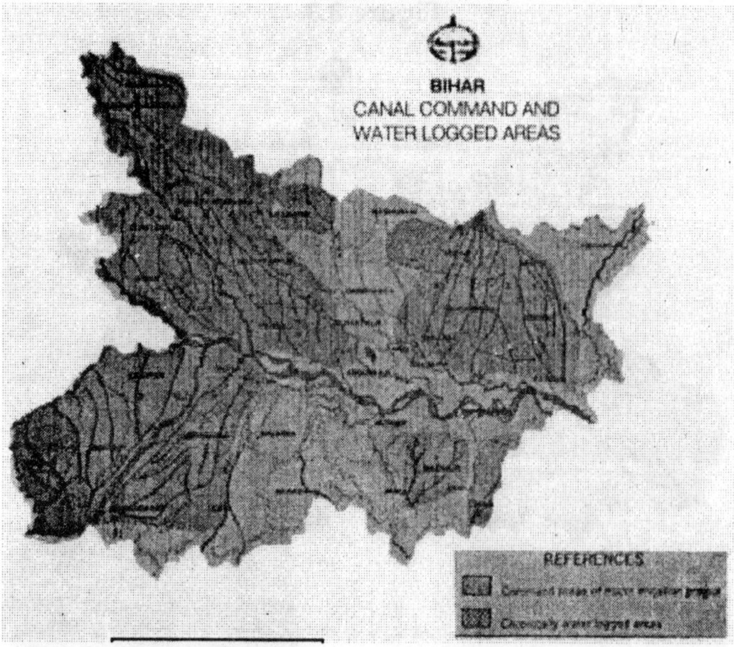

The soil temperature regime is hypothermic and the soil moisture regime is ustic except in the eastern Himalayan foothills and piedmont zone where it is udic. In low lying areas where water stagnates during rainy season qualify for aquic moisture regime. Based on the variability in rainfall (P), potential evapotranspiration (PET), actual evaop-transpiration (AE) relation between P and PET and AE and length of growing period (LDP) for normal cropping system, the entire terrain of Bihar state forms different ecological situation.

The agricultural economy of the state is primarily a food crop economy and more than 90 per cent of the cropped areas are devoted to food grains. Agricultural productivity is rather low because of adoption of traditional methods of cultivation and use of local varieties by the small and marginal farmers. Flood is the annual feature in Bihar. The productivity of important crops is given below (Table 3.2).

Table 3.2: Productivity of Principal Crops in the State

Crops	*Average yield (g/ha)*						*Potential*
	1980	*1985*	*1990*	*1995*	*2001*	*2005*	*yield (q/ha)*
Rice	10.29	11.99	12.89	13.71	14.89	10.75	30.26
Wheat	13.30	15.10	18.80	2.08	21.47	16.09	30.52
Maize	10.21	13.9	16.4	19.42	24.13	23.86	40.50
Gram	7.2	9.0	9.8	10.5	9.48	8.35	13.80
Arhar	12.0	13.8	13.3	12.0	13.50	12.35	18.30
Lentil	6.8	8.0	7.9	9.6	8.60	-	13.80
Rapeseed/mustard	5.2	7.8	8.1	7.7	9.33	9.89	18.50
Sugarcane	340.8	343.9	422.0	414.4	424.24	394.6	627.80
Potato	91.7	92.0	91.2	90.5	96.9	136.4	235.00

Source: Kisan Diary 2007 RAU, Pusa.

Rice is grown in "kharif, summer and boro" seasons. In Bihar, rice is main staple food. Its percapita monthly consumption has declined during last 30 years, but still it constitutes major portion of diet. Areas under rice increased in eighties, which stagnated during nineties and even in first give years of 21st century. However production has continuously increased during last 20 years due to increase in rice productivity from 1029 kg/ha in 1980-81 to 1660 kg/ha in 2005-06 (Table 3.3).

Table 3.3: Projected Rice Area, Production and Productivity in Different Agro-Climatic Zones of Bihar (2000-01 to 2025-26)

Projected Rice Area *(In lakh ha)*

Year	*Zone-I*	*Zone-II*	*Zone-III a*	*Zone-III b*	*Bihar*
2000-01	13.62	8.17	3.02	13.71	38.52
2005-06	13.32	8.48	2.92	13.79	38.51
2010-11	13.03	8.80	2.83	13.85	38.51
2015-16	12.75	9.13	2.83	13.92	38.63
2020-21	12.47	9.47	2.83	13.49	38.26
2025-26	12.20	9.83	2.83	14.06	38.92

Projected Rice Production

Year	*Zone-I*	*Zone-II*	*Zone-III a*	*Zone-III b*	*Bihar*
2005-06	21.84	12.66	3.83	25.70	64.02
2010-11	26.66	16.31	5.16	29.83	77.96
2015-16	31.62	20.47	7.00	45.11	104.20
2020-21	38.04	26.04	8.56	51.83	124.47
2025-26	45.77	31.51	9.90	63.27	150.45

Projected Rice Productivity

Year	*Zone-I*	*Zone-II*	*Zone-III a*	*Zone-III b*	*Bihar*
2000-01	1.60	1.55	1.26	1.87	1.66
2005-06	1.64	1.49	1.31	1.86	1.66
2010-11	2.05	1.85	1.82	2.15	2.02
2015-16	2.48	2.25	2.47	3.24	2.69
2020-21	3.05	2.75	3.02	3.84	3.25
2025-26	3.75	3.21	3.50	4.50	3.86

3.5 Madhubani

Bihar has three broad climate zone, Zone-I, Zone-II and Zone-III. Zone II is a traditional boro rice growing region where soil gorund water availability and temperature are favorable for the crop. Through Zone-I has similar soil profile but low temperature is the main constraints. In zone-III A and III B only selected pockets are suitable for boro cultivation. In zone I and zone II temperature drop as low as below 5°C and it is the serious constraints at seedling stage.

The Madhubani district fall under North Eastern Plain Zone of Agro-climatic zone II. The District is situated between 85°43` and 86°42` north latitude and 25°59` and 26°39` east longitude. The geographical area of the district is 3501 sq. km. The average rainfall in 1273.2mm. The district of Madhubani was carved out of the old Darbhanga district in the year 1972 as a result of reorganization of the districts in the state. This was formerly the northern subdivision of Darbhanga district. Bounded on the north by a hill region of Nepal and extending to the border of its parent district Darbhanag in the south, Sitamarhi in the west and Supaul in the east, Madhubani fairly represents the culture of the territory once known as Mithila and the district has maintained a distinct individuality of its own. The district has five sub-divisions and twenty one blocks. The blocks are—Andharathari, Laukahi, Babubarhi, Basopatti, Benipatti, Bisfi, Harlakhi, Ghoghardiha, Jainagar, Jhanjharpur, Kaluahi, Khajauli, Khutauna, Ladania, Lakhanaur, Madhepur, Madhwapur, Pandaul, Phulparas, Rahika and Rajnagar. Madhubani district is one of the thirty-eight districts of Bihar state, where Madhubani is the district headquarters. Madhubani is a District that had total population of 3,575,281 second lowest in Bihar only to Kaimur (District) on Urbanization Rate. It ranks 5th highest on Sex Ratio of any District in Bihar. It is second highest in Bihar only to Nawada (District) on Urban SC sex Ratio. It is second highest in Bihar only to Khagaria (District) on Under 6 ST Sex Ratio is most suitable for research on Boro Rice.

Figure 3.5

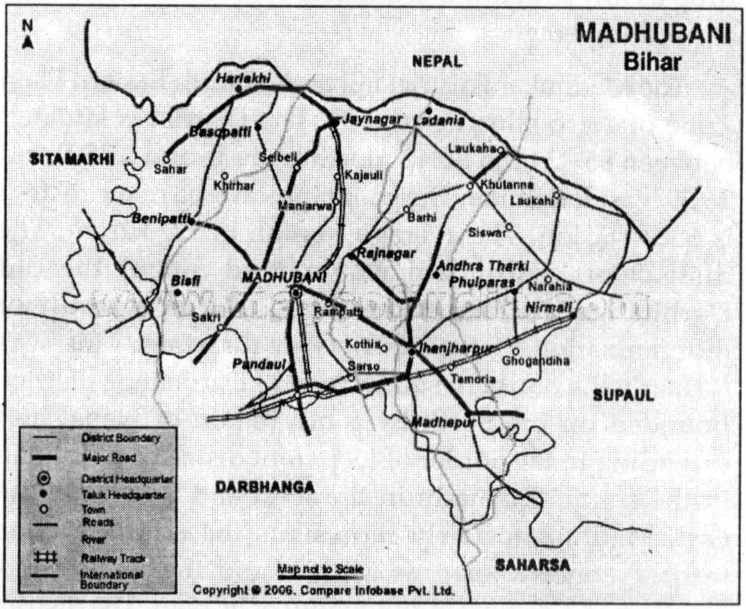

3.6 Administrative Division

The district consists of five sub-divisions and 21 blocks (table 3.7).

Administrative Division

Sl.No.	Particulars	Number
1.	No. of Sub-division	5
2.	No. of Blocks	21

3.	No. of towns	5
4.	No. of Panchayats	399
5.	No. of Villages	1111
6.	No. of Police Stations	18

Source: Census 2001.

3.6.1 River

The rivers that flow in Madhubani District and near to the district are the Kosi river, Kamla Balan, Bhutahi Balan, Suragwe, Tilyuga, Jeevach and Dhosh. The district intersected by Kamla and Bhutahi Balaan rivers is flourishing every year with its farming endeavors and artistic richness.

3.6.2 Rainfall Pattern

The normal rainfall of the district is 1273.2mm percentage. Nearly 85.95 percent annual rainfall is received during June to September.

3.6.3 Source Wise Area under Irrigation

Percentage of irrigation has increased form 37.92 is 1985-86 to 54.71 in 1995-96 (Table 3.4). Major proportion of area under irrigation was by private tube well. During the mentioned period share of irrigation by state tube well and tanks, rivers and other sources has decreased and that of by private tube well has increased substantially.

Table 3.4: Source Wise Area Under Irrigation

Sl. No.	Source	1985-86		1995-96	
		Irrigated area (ha)	% of net sown area	Irrigated area (ha)	% of net sown area
1.	State tube well	7416	4.19	637	.036
2.	Private tube well	50495	28.55	91622	51.12
3.	Tanks, rivers and other sources	6165	5.18	5616	3.23
	Total	67076	37.92	97875	54.71

Source: Lead Bank Officer, Dist. Planning Cell, RAU, Pusa.

3.6.4 Cropping Pattern

Cropping pattern of the district is mainly food grain oriented. Average of 7 years of area under major crops indicated total area under food grains was 88.86 per cent of gross cropped area in which highest area was under paddy (39.30 per cent) followed by wheat (23.79),maize (5.91 per cent) and others crops (19.86 per cent).

3.6.5 Fertilizer Consumption

Fertilizer consumption in term of NPK is given in Appendix-IV Average annual consumption of the district was 21038 matric tonnes, of which kharif and rabi season account for 34.10 and 65.90 per cent, respectively. Per hectare fertilizer consumption in term of NPK was worked out 80.22 kilograms. Madhubani and Darbhanga district recorded for Gautam performance was 8-9 to 7.0 t/ha (Table 3.5). Due to drought or flood singly or in combination rice production in Bihar has been affected. Boro rice cultivation on large scale in prospective area will ensure higher rice production and stabilize the rice productivity in the state at the same time it will compensate the loss of Kharif rice.

Table 3.5: Performance of Gautam in Front Line Demonstration During 1994-95 to 1997-98

Year	*Area (ha)*	*Block*	*District*	*Av. Yield (t/ha)*	*Remarks*
1994-95	8.0	Madhepur	Madhubani	8.4	This variety is
1995-96	8.0	Madhepur	Madhubani	8.9	now most
1996-97	8.0	Lakhnaur	Madhubani	7.5	popular in boro
1997-98	6.5	Biraul	Darbhanga	7.0	season.

Table 3.6: Details of Boro Rice Variety Gautam, Grown in Israin, Chaur, Madhepur, and Daliara Chaurs in Lakhnaur block of Madhubani District During 1999-2000

Sl. No.	*Name of Village*	*No. of farmers*	*Area (ha)*	*Total yield*	*Yield (t/ha)*
1.	Nirmala	120	25.0	200.0	8.0
2.	Kasayam	20	7.0	57.0	8.2
3.	Khairi	14	8.0	62.40	7.8
4.	Gunakarpur	07	5.40	40.50	7.5
5.	Prasad	33	11.0	77.0	7.0
6.	Pachahi	35	11.0	82.5	7.5
7.	Kataki	69	19.0	152.0	8.0
8.	Muth Bhagawanpur	05	02.0	16.4	8.2
9.	Birpur	15	06.0	48.0	8.0
10.	Sikaria	12	04.0	28.0	7.0
11.	Ram Chandra	38	15.0	115.0	7.7
12.	Pure	36	14.0	100.8	7.2
13.	Bahera	9	2.0	13.6	6.8
14.	Sunderpur	9	2.50	19.0	7.2
15.	Dashen amrupi	4	1.25	10.0	8.0

16.	Biharpur Mushahari	11	1.25	10.0	8.0
17.	Madhupur	15	6.00	46.8	7.8
18.	Khajuri	11	4.00	30.00	7.5
19.	Chunni	4	1.00	7.00	7.0
20.	Khajura	2	.075	5.4	7.2
21.	Karhara	1	0.50	4.4	8.8
22.	Manmohan	1	0.25	2.1	8.4
23.	Nawada	2	0.75	6.1	8.1
24.	Telgaraha	2	0.75	6.0	8.0
25.	Senaborsa	20	7.00	50.4	7.2
26.	Barhmotar	2	0.75	5.5	7.3
27.	Bochahi	8	2.25	14.6	6.5
28.	Pochmani	5	1.50	10.2	6.8
29.	Mavi	2	0.50	3.8	7.5
30.	Beleurcha	1	0.40	3.0	7.6
31.	Loaf	65	20	136.0	6.8
	Grand Total	575	180.0	1365.5	AV. 7.7

3.7 Future Prospect of Boro Rice in Bihar

There are vast waterlogged area and chaur land in Bihar, where wheat crop can not be grown due to water logging. These area are potential for boro in real sense, with viability of supplemental irrigation at the late stage of boro crop. At present the crop is mostly cultivated in north eastern part of Bihar in the districts of Saharsa, Purnia and Katihar, where climatic condition are favorable. Its cultivation, however, did not spread beyond this region mainly because of the unavaibility of cold tolerant boro rice varieties, because the low temperature is the main constraints in other region. With release of cold tolerant varieties Gautam, Richharia, Dhanlaxmi,

Saroj and Prabhat its cultivation has been expanded in other region of state such as Darbhanga and Madhubani districts. A new prospective area in the district of Sitamarhi have been recently identified use of low cost polytunel as an alternative approach to save boro rice nursery from cold during winter is an effective technology. Combination of both cold tolerant varietals technology and polytunel technology will be most effective in wide expansion of boro rice area in Bihar.

4

Findings and Discussion

The results of the research are presented and discussed in this chapter after appropriate statistical analysis. The respondents were classified into marginal, small and medium farmers for in-depth analysis. The findings in six sections.

In the first sections, differential characteristics of marginal, small and medium farmers were studied with respect to selected agro-economic, socio-psychological and communicational variables. In the second section deals with technological gap in the adoption of various components of boro rice production technology. Third section training needs of marginal, small and medium farmers in the ten main areas and forty-six areas of training in relation to boro rice production technologies were studied. The training need perception technologies ere studied. The training need perception of Extension Officers were also included in this section. Section four deals with the association of selected independent variables with the training need of farmer in boro rice cultivation along with multiple regression analysis showing the prediction potency of the variables in training need. The preferential choice of different categories of farmers in respect of various components related to boro rice training programme has been discussed in the fifth section. The last section deals with the training strategy for accelerating winter boro rice production based on the findings of the present investigation.

SECTION – I

4.1 Differential Characteristics of the Marginal, Small and Medium Farmers

It is commonly believed that marginal, small and medium farmers differ in some characteristics. The personal characteristics of the farmers, their resource status, farm situation and level of utilization of information jointly play an important role in decision making towards adopting an innovation. At the outset, it was intended to have some idea of the differential characteristics of marginal, small and medium farmers with respect to selected agro-economic (5), socio-psychological (7) and communicational (2) variables. For this purpose, the mean value of independent variables was worked out and is presented in Table 4.1. The significance of difference between the mean values of selected variables of marginal and small, marginal and medium and small and medium farmers were tested by 't' test and are presented in Table 4.3.

4.1.1 Agro-Economic Variables

Farm Size: The average farm size of the marginal, small and medium farmers were 0.55, 1.55 and 3.47 hectares respectively. The computed 't' value indicated that the marginal farmers had significantly smaller farm size in comparison to both the small and medium farmers. The small farmers had also significantly small farm size in comparison to the medium farmers.

4.1.2 Irrigation Index

The mean irrigation index of the marginal, small and medium farmers were 60.47, 82.15 and 80.18 per cent respectively. It is obvious that small farmers had higher irrigation index in comparison to medium and marginal farmers. The computed 't' values indicated that I comparison to both the small and medium farmers, the marginal farmers had significantly lasser percentage of area under irrigation. The difference in the percentage of area under irrigation of the small and medium farmers was non-significant.

4.1.3 Cropping Intensity

The mean cropping intensity of the marginal, small and medium farmers were 152.53, 168.47 and 184.75 per cent respectively. The computed 't' values indicated that marginal farmers had significantly lower cropping intensity as compared to both small and medium farmers. Small farmers had also significantly lower cropping intensity in comparison to medium farmers.

Table 4.1: Mean Values of Independent Variables

Characteristic	*Unit*	*Mean Values*		
		Marginal farmers (N = 119)	*Small farmers (N = 54)*	*Medium farmers (N = 27)*
AGRO-ECONOMIC				
Farm Size	Hectare	0.553	1.548	3.468
Irrigation index	Percent	60.472	82.149	80.175

Cropping Intensity	Percent	152.533	168.469	184.747
Farm Mechanization	Score	23.101	26.333	29.963
Technological gap	Percent	53.80	49.07	44.50
SOCIO-PSYCHOLOGICAL				
Age of the farmers	Score	42.748	37.444	41.470
Education of the farmer	Score	2.109	2.278	3.037
Family size	Number	7.605	7.963	6.815
Knowledge level	Percent	32.803	45.244	43.4000
Risk orientation	Score	27.723	30.333	25.037
Economic motivation	Score	3.445	4.845	4.211
Market orientation	Score	21.101	19.111	16.630
COMMUNICATION				
Utilization of mass media	Score	11.899	17.769	19.074
Utilization of personal cosmopolite	Score	11.613	14.500	16.111

Table 4.2: Significance of Differences in the Mean Values of Independent Variables

Characteristics	*'t' Values*		
	Marginal and small farmers	*Marginal and medium farmers*	*Small and medium farmers*
AGRO-ECONOMIC			
Farm size	20.981**	15.132**	9.835**
Irrigation index	12.183**	11.560**	1.205
Cropping intensity	5.837**	9.734**	4.190**
Farm mechanization	3.030**	4.048**	1.972**
Technological gap	6.281**	9.441**	2.802**
SOCIO-PSYCHOLOGICAL			
Age of the farmers	1.246	2.422**	1.389
Education of the farmers	0.796	3.533**	2.683*
Family size	0.484	1.413	1.667

Knowledge level	5.703**	6.854**	1.221
Risk orientation	3.208**	1.781	1.670
Economic motivation	4.666**	1.854	3.436**
Market orientation	3.117**	4.583**	1.369
COMMUNICATIONAL			
Utilization of mass media	19.814**	20.185**	1.023
Utilization of personal cosmopolite	4.293**	4.943**	1.670

4.1.4 Farm Mechanization

The mean farm mechanization scores of the marginal, small and medium farmers were 23.10, 26.33 and 29.96 respectively. The computed 't' values indicated that in comparison to both small and medium farmers the marginal farmers had significantly low level of use of farm machinery. There was no significant difference in the level of use of farm machinery by the small and medium farmers.

4.1.5 Technological Gap

The mean technological gap of the marginal, small and medium farmers were 53.80, 49.07 and 44.50 per cent respectively. The computed 't' values indicated that in comparison to both small and medium farmers, marginal farmers had significantly higher technological gap. The small farmers had also significantly higher technological gap in comparison to medium farmers.

4.1.6 Socio-Psychological Variables

Age of the Farmers: The mean age of the marginal, small and medium farmers were 42, 37and 41 years respectively. The computed 't' values indicated that there was no significant difference between marginal and small and the small medium farmers with respect to their age. The age of the marginal farmers was significantly higher than medium farmers.

4.1.6.1 Education of the Farmer

The mean education scores of the marginal, small and medium farmers were 2.11, 2.28 and 3.04, respectively. The computed 't' values indicated that in comparison to both marginal and small farmers, medium farmers had significantly higher education level. The level of education of the marginal and small farmer was not significantly different.

4.1.6.2 Family Size

The mean number of members in the family of the marginal, small and medium farmers were 7, 7 and 6 respectively. By the computed 't' values for the three categories of farmers, it could be inferred that there was no significant difference between the marginal and small, the marginal and medium and the small and medium farmers with respect to family size.

4.1.6.3 Knowledge Level

The mean score of level of knowledge about boro rice technology of the marginal, small and medium farmers were 32.80, 45.24 and 43.40 respectively. The computed 't' values indicated that in comparison to small and medium farmers, the marginal farmers had significantly lower knowledge about boro rice production technology. However, there was no significant difference between the small and medium farmers in respect of their knowledge about boro rice technology.

4.1.6.4 Risk Orientation

The risk orientation scores of the marginal, small and medium farmers were 27.72, 30.33 and 25.04, respectively. The computed 't' values indicated that the risk orientation of the marginal farmers was significantly lower in comparison to small farmers. The marginal farmers were less orientated to take risk in farming. The risk orientation of medium farmers was not significantly different in comparison to both marginal and small farmers.

4.1.6.5 Economic Motivation

The mean economic motivation scores of the marginal, small and medium farmers were 3.44, 4.84 and 4.21, respectively. The computed 't' values indicated that in comparison to both marginal and medium farmers, the small farmers had significantly higher economic motivation. However, the economic motivation of the marginal and medium farmers was not significantly different.

4.1.6.6 Market Orientation

The mean scores on market orientation of the marginal, small and medium farmers were 21.10, 19.11 and 16.63, respectively. The computed 't' values indicated that in comparison to both small and medium farmers, the marginal farmers had significantly more market orientation. The difference in market orientation of the small and medium farmers was not significant.

4.1.7 Communicational Variables

4.1.8 Utilization of Mass Media

The mean scores of utilization of mass media of the marginal, small and medium farmers were 11.90, 17.77 and 19.07, respectively. The computed 't' values indicated that in comparison to both the small and medium farmers, the marginal farmers had significantly lesser utilization of mass media. There was no significant difference in the utilization of mass media by the small and the medium farmers.

4.1.9 Utilization of Personal Cosmopolite

The mean score of utilization of personal cosmopolite sources of information of marginal, small and medium farmers were 11.61, 14.50 and 16.11, respectively. The computed 't' values indicated that in comparison to both the small and medium farmers, the marginal farmers has significantly lesser utilization of personal cosmopolite sources. The utilization of personal cosmopolite sources of information by the small and the medium farmers was not significantly different.

On the basis of the data in Table 4.1 and 4.2, comparative profiles of the marginal, small and medium farmers are presented in Table 4.3. The important points which emerged out of the study of the differential characteristics of the marginal, small and medium farmers are as under:

4.1.9.1 In comparison to both the small and medium farmers, the marginal farmers had significantly smaller farm size, lower irrigation index, lower farm mechanization, more technological gap, lesser knowledge about boro rice technology, more market orientation, less utilization of mass media and less utilization of personal cosmopolite sources.

Table 4.3: Comparison of the Profiles of Marginal, Small and Medium Farmers

Characteristics	*'t' Values*		
	Marginal and small farmers	*Marginal and medium farmers*	*Small and medium farmers*
AGRO-ECONOMIC			
Farm size	Small	Small	Small
Irrigation index	Low	Low	Low
Cropping intensity	Low	Low	Low
Farm mechanization	Low	Low	NSD*
Technological gap	More	More	More
SOCIO-PSYCHOLOGICAL			
Age of the farmers	NSD*	High	NSD*
Education of the farmers	NSD*	Low	Low
Family size	NSD*	NSD*	NSD*
Knowledge level	Less	Less	NDS
Risk orientation	Low	NSD*	NSD*
Economic motivation	Low	NSD*	High
Market orientation	More	More	NSD*
COMMUNICATIONAL			
Utilization of mass media	Less	Less	NSD*
Utilization of personal cosmopolite	Less	Less	NSD*

* NSD = No Significant Difference

4.1.9.2 In comparison to medium farmers, the marginal farmers had significantly higher age and lower educational level. In comparison to the medium farmers, the small farmers had significantly smaller farm size, lower cropping intensity, more technological gap and higher economic motivation.

4.1.9.3 No significant difference was found between the small and medium farmers with respect to irrigation index, farm mechanization, age of the farmers, knowledge level, risk orientation, market orientation, utilization of mass media and utilization of personal cosmopolite sources of information.

4.1.9.4 There was no significant difference between the marginal, small and medium farmers with respect to family size.

SECTION – II

4.2 Extent of Technological Gap in Boro Rice Cultivation

Improved boro rice production technologies in the form of package of practices are developed by the scientists and recommended to the farming community for adoption. The objective is to increase the production and productivity of the crop. In spite of numerous and concerted efforts of agricultural scientists, thee still exists a wide gap between the recommended technologies and the adoption of these technologies in the field by the farmers. In Bihar, were occupies a pride place among rabi cereals, the average per hectare yield is about 2 to 5 tonnes in farmer's filed in comparison to 6-9 tonnes in demonstration plots which further exhibit substantial possibility to increase the productivity.

The yield of boro rice is dictated by the extent of adoption of various components of production technologies. Therefore, it is necessary to reduce the technological gap in the field of the boro rice growing farmers of Bihar and other boro rice production states for increasing the crop yields at desired level. The results have been presented in Table 4.4.

Table 4.4: Extent of Technology Gap of Farmers in Boro Rice Cultivation

Sl. No.	*Practices of boro rice cultivation*	*Maximum attainable scores*	*Mean scores obtained*	*Gap in score*	*Gap in %age*	*Ranks*
1.	Improved seed and their treatment	14	5.01	8.99	64.23	I
2.	Sowing methods	6	2.66	3.34	55.67	III
3.	Fertilizer application	24	10.07	13.93	58.05	II
4.	Irrigation	6	2.79	3.21	53.50	IV
5.	Pest control	21	9.83	11.17	53.19	V
6.	Weed control	16	10.25	5.75	35.92	VII
7.	Post harvest technology	4	2.49	1.51	37.82	VI

The study of technological gap in the knowledge and adoption level of recommended package of practices by the farmers clearly revealed almost similar trend in case of fertilizer application by small medium and marginal farmers in general. The reason is obvious as most of farmers are not aware of balanced and timely application of fertilizer. The less knowledge of requirement of micro nutrient and integrated nutrient management also add this fact. There is a wide gap in case in irrigation also in case of small as well as medium farmers the reason being and inadequate knowledge of interaction between nutrients and water application. Further farmers also have less knowledge about critical growth phases and appropriate time of irrigation application for higher productivity.

The marginal farmers having less knowledge of the merits of improved seed and their sources of availability showed widest gap in this case. The lowest gap observed in post-harvest technology in case of small, medium as well as in general farmers was due to its more mechanical adoption. Further, it is not felt essential, as it does not add to the productivity.

SECTION – III

4.3 Perception of Training Need of Farmers in Respect of Boro Rice Production

The relative training need perception of marginal, small and medium farmers in relation to boro rice production technology in the ten main areas are presented in this section. Further, attempts have also been made to ascertain the perception of farmers in the forty six sub-areas under the main area of training.

4.3.1 Perception of Training Need of Farmers in the Main Areas of Boro Rice Technology

The study reveals that improved cultural practices emerged as the topmost priority training area in boro rice cultivation. The marginal farmers gave vital importance to use hybriding rice technology it second in priority. The component integrated nutrient management and water management have been given the ranks at 3rd and 4th position in the priority. Integrated pest management, seed treatment and land preparation have been ranked respectively at 5th, 6th and 7th position by the marginal farmers. Post harvest technology, cropping system and weed control were accorded the last of the three choices in training priority area by the farmers.

The study also reveals that small farmer have similar view regarding the training need priority in improved cultural practices in boro rice cultivation as the most needed area for training followed by 2nd priority in integrated nutrient management. The table further reveals that use of hybrid rice is the third priority area for training by small farmers. Water management, integrated pest management and post harvest technology have been ranked as 4th, 5th and 6th positions respectively in training priority by the small farmers. Cropping system, weed control and land preparation have been accorded the last of the three choice in making pattern by the small farmers in boro rice training.

4.3.2 Training Need of Farmers in the Sub-Area of Boro Rice Technology

. The relative need for training need for training in each sub-area of training component as perceived by the three categories of farmers are presented according to the main area of training.

4.3.3 Land Preparation

Table 4.5 shows that all the three categories of farmers have almost the same view regarding the training need in reduced tillage for land preparation as the most needed area for training. Whereas, small and medium farmers have shown similar priority in training need in tillage equipments and normal tillage by giving 2nd and 3rd rank respectively.

Table 4.5: Training Need of the Farmers in the Sub-Area of Land Preparation in Boro Rice

Sl. No.	*Sub-areas of training*	*Marginal farmers (N=119)*		*Small farmers (N=54)*		*Medium farmers (N=27)*		*Pooled (N=200)*	
		ACS	*Rank*	*ACS*	*Rank*	*ACS*	*Rank*	*ACS*	*Rank*
1.	Normal tillage	100.33	II	42.00	III	20.66	III	54.33	II
2.	Reduced tillage	107.33	I	52.66	I	24.33	I	61.44	I
3.	Tillage equipments	90.33	III	44.33	II	23.00	II	52.56	II

4.3.4 Weed Control

Table 4.6 reveals that perception of training need is similar in small and medium farmers in different training sub-areas of weed control. However, marginal farmers have different perception in sub-areas of weed management except weed control by manual, chemical and mechanical means, which was accorded Ist rank by all the three categories of farmers. Knowledge of preventive method was accorded 2nd priority by small and medium farmers. Whereas, marginal farmers assigned 2nd priority to identification of weeds followed by 4th priority given by small and medium farmers in the same area. Knowledge of preventive method was accorded 3rd rank by the marginal farmers, whereas, small and medium farmers have given 3rd priority for training in weed control under different cropping systems. All the three categories of farmers perceived least training need area in availability and economic of herbicide.

The pooled data reveals that weed control by manual, chemical and mechanical, knowledge of preventive method and identification of weeds have been given Ist, 2nd and 3rd priority followed by preparation of spray volume, weed control under different cropping systems, and availability and economics of herbicide.

Table 4.6: Training Need of the Farmers in the Sub Area of Weed Control in Cultivation of Boro Rice

Sl. No.	*Sub-areas of training*	*Marginal farmers (N=119)*		*Small farmers (N=54)*		*Medium farmers (N=27)*		*Pooled (N=200)*	
		ACS	*Rank*	*ACS*	*Rank*	*ACS*	*Rank*	*ACS*	*Rank*
1.	Identification of weed	92.00	II	37.66	IV	18.33	IV	49.33	III
2.	Knowledge of preventive method	89.33	III	41.66	II	21.33	II	50.78	II
3.	Weed control by manual, chemical and mechanical	105.33	I	52.33	I	25.33	I	61.00	I
4.	Preparation of spray volume	85.00	IV	37.33	V	17.66	V	46.66	IV
5.	Availability and economic of herbicide	67.33	VI	31.66	VI	17.00	VI	38.67	VI
6.	Weed control under different cropping system	70.33	V	41.33	III	20.00	III	43.88	V

4.3.5 Water Management

Table 4.7 reveals that all the three categories of farmers have given similar perception for training in time of irrigation and accorded it as the most needed area of training followed by 2nd priority in water requirement of Rice based crop sequence by small and medium farmers. Whereas, marginal farmers have accorded 3rd rank in this area. Depth of irrigation water have shown 2n priority by the marginal farmers. While this component have been given 3rd rank by the small and medium farmers. Water requirement of rice based intercropping system and method of irrigation have been accorded least priority for training by all the three categories of farmers.

Table 4.7: Training Need of the Farmers in the Sub Area of Water Management in Cultivation of Boro Rice

Sl. No.	*Sub-areas of training*	*Marginal farmers (N=119)*		*Small farmers (N=54)*		*Medium farmers (N=27)*		*Pooled (N=200)*	
		ACS	*Rank*	*ACS*	*Rank*	*ACS*	*Rank*	*ACS*	*Rank*
1.	Time of irrigation	114.33	I	50.00	I	25.00	I	63.11	I
2.	Method of irrigation	69.00	V	32.33	V	16.00	V	39.11	V
3.	Depth of irrigation water	100.33	II	38.00	III	19.66	III	52.66	II
4.	Water requirement of maize based intercrop	78.66	IV	37.33	IV	19.33	IV	45.10	IV
5.	Water requirement of maize based crop sequence	80.66	III	49.00	II	22.66	II	50.77	III

The pooled data reveals that time of irrigation and depth of water have been reckoned as the most needed priority area of for training followed by water requirement of boro rice based crop sequences. Water requirement of boro rice based intercropping and method of irrigation was preferred as the least needed training area and all the categories of farmers assigned last priority for training in this area.

SECTION – IV

4.4 Association of Training Needs of Farmers with Selected Agro-Economic, Socio-Psychological and Communicational Variables

Training needs of farmers for increasing boro rice productivity was dependent variable (Y) and five agro-economic variables i.e. farm size (X1), irrigation index (X2), cropping intensity (X3), farm mechanization (X4) and technological gap (X5), seven socio-psychological variables like age of farmers (X6) education of the farmers (X7), family size (X8), knowledge level (X9), risk orientation (X10), economic motivation (X11) and market orientation (X12) and two communicational variables were utilization of mass media (X13) and utilization of personal cosmopolite (X14). Based on the review of literature and discussion with scientist and extension workers variables were selected for this study. The coefficient of correlation of training need (Y) with the selected independent variables of different categories of farmers was worked out and presented in Table 4.8.

The training need of farmers in boro rice production technology may be affected by various agro-economic, socio-psychological and communicational variable of the respondent. Table reveals the nature and extent of association of independent variables with the training need perception in boro rice cultivation with marginal, small and medium farmers. It is clear from the table that same set of variables behaved differently in exhibiting its association with the training need perception in the three categories of farmers. A general view of association of different variables revealed that cropping intensity, economic motivation and utilization of personnel cosmopolite were the only three variables which displayed positive and significant association while only one variable namely age of the farmers was found significant but negative association with the training need irrespective of categories.

Table 4.8: Association between Selected Independent Variables with Training Need of Farmers in Boro Rice Technology

Sl. No.	*Independent Variables*	*'t' Values*			
		Marginal farmers (N=119)	*Small farmers (N=54)*	*Medium farmers (N=27)*	*Pooled (N=200)*
A.	AGRO-ECONOMIC				
X1	Farm size	0.732**	0.076	0.300	0.213*
X2	Irrigation index	0.075	0.179	0.297	0.032
X3	Cropping intensity	0.694**	0.828**	0.873**	0.643**
X4	Farm Mechanization	0.023	-0.149	0.896**	0.129
X5	Technological gap	-0.159	0.365**	-0.096	0.023
B.	SOCIO-PSYCHOLOGICAL				
X6	Age of the farmers	-0.670**	-0.740**	0.517**	-0.669**
X7	Education of the farmers	0.730**	0.116	-0.058	0.475**
X8	Family size	0.066	-.578**	-0.324	-0.183
X9	Knowledge level	0.087	-0.147	-0.197	0.059

X10 Risk orientation	0.129	0.027	0.951**	0.246*
X11 Economic motivation	0.743**	0.887**	0.532**	0.731**
X12 Market orientation	-0.006	0.156	0.977**	0.098
COMMUNICATIONAL				
X13 Utilization of mass media	0.101	0.864**	0.950**	0.287**
X14 Utilization of personal cosmopolite	0.474**	0.008	0.921**	0.405**

** Significant at 1 per cent level of probability.

* Significant at 5 per cent level of probability.

4.4.1 Association between the Training Needs of the Marginal Farmers and the Independent Variables

4.4.1.1 Agro-Economic Variables

Table 4.8 revealed that the training need of the marginal farmers was positively and significantly associated with their farm size and cropping intensity at 1 percent level. The results indicated that in case of marginal farmers, the increase in the training need was associated with increase in the percentage of farm size and cropping intensity. For the marginal farmers, it is due to practiced intensive cultivation on their field and desire to more information about farming operation for increasing productivity. Therefore, farmers have bigger farm size possessed more training need.

The training need of the marginal farmers was not significantly correlated with their irrigation index, farm mechanization and technological gap. The association of the variables with training need of the marginal farmers was not revealed, probably because the marginal farmer had lesser percentage of area under irrigation, low level of use of farm machinery and low level of adoption of improved maize technology on their farm. Therefore, these variables were not relevant in determining the training needs.

4.4.1.2 Socio-Psychological Variables

Table 4.8 revealed that training need of the marginal farmers was positively and significantly associated with their education, and economic motivation at 1 per cent level of probability. The training need was negatively and significantly correlated with age of the farmers at 1 per cent level of probability. The results indicated that in case of marginal farmers, the increase in the training need was associated with increase in their education and also increase in economic motivation. For the marginal farmers greater was the age of the farmers, the lesser was the incentive for the training need.

The training need of the marginal farmers was not significantly associated with their family size, risk orientation and market orientation. The results indicated that the variable family size, knowledge level, risk orientation and market orientation were not relevant in determining the training need of marginal farmers. Probably because being marginal farmers they were less orientated to take risk in farming and less market oriented.

4.4.1.3 Communicational Variables

Table 4.8 revealed that the training need of marginal farmers was positively and significantly associated with utilization of personal cosmopolite at 1 per cent level of probability.

The results indicated that in case of marginal farmers, the increase in the training need was associated with their more utilization of cosmopolite sources of information. Incidentally, utilization personal cosmopolite sources of information provided opportunity for personal face to face communication of the marginal farmers with the change agents.

The association of the above variables with the training need of the marginal farmers was not evident, probably because the marginal farmers had low utilization of mass media for getting information about improved boro rice technology.

SECTION – V

4.5 Opinion of the Farmers Regarding Different Components of Training Programme

After ascertaining the training needs of farmers with respect to boro rice production technology, their association of different factors along with prediction potencies in effecting the training needs and the direct and indirect effects of various factors on the training need, it was thought essential to know the farmers opinion with respect to venue for training, period of training, duration of training, size of training, different methods of training and follow up activities. The component wise findings were analyzed and have been presented below:

4.5.1 Venue and Time Training

The respondents were given option to give their potential choice for different and timings venues of training. Their response pattern has been indicated in table 4.9 & 4.10.

Table 4.9 indicates that organization of training programme at the village level and at the demonstration plot was most and moderately preferred by the marginal farmers as they assigned Ist and IInd rank respectively. Training at the Panchayat level and at the block level was given third and fourth choice by the marginal farmers. Venue at the university/college campus for training was given a last preferential choice by the marginal farmers. Organization of training programme at the village level and at the demonstration plot was given Ist and IInd choice by the small farmers followed by training programme at the university/college campus and at the Panchayat level. Venue at the block level for training was given a last preferential choice by the farmers.

Table 4.9: Farmers Option for Training Venue

Sl. No.	*Venue for training*	*Marginal farmers (N=119)*		*Small farmers (N=54)*		*Medium farmers (N=27)*		*Pooled sample farmers (N=200)*	
		Mean score	*Rank*	*Mean score*	*Rank*	*Mean score*	*Rank*	*Mean score*	*Rank*
1.	Village level	3.41	I	3.29	I	3.33	II	3.34	I
2.	Panchayat level	2.93	III	2.66	IV	2.85	V	2.81	IV
3.	Block level	2.71	IV	2.44	V	3.07	IV	2.74	V
4.	University/ College campus	2.67	V	2.70	III	3.48	I	2.95	III
5.	Demonstration plot	3.31	II	2.88	II	3.25	III	3.14	II

Organization of training programme at the university/college campus and at the village level was given Ist and IInd choice respectively by the medium farmers followed by training programme at the demonstration plot, and at the block level by assigning 3rd and 4th choice respectively. Venue at the Panchayat level for training was given a last preferential choice. Pooled analysis shows that organization of training programme at the village level was most preferred and was given Ist rank by the farmers.

After analyzing the venue for training, an effort was made to ascertain the differential opinion of the farmer regarding timing for organizing training camp with respect to boro rice production technology. The respondent were given option to give their preferential choice for the time of training. Their response pattern was presented in Table 4.10.

Table 4.10: Farmers Option for Time of Training

Sl. No.	*Time for training*	*Marginal farmers (N=119)*		*Small farmers (N=54)*		*Medium farmers (N=27)*		*Pooled sample farmers (N=200)*	
		Mean score	*Rank*	*Mean score*	*Rank*	*Mean score*	*Rank*	*Mean score*	*Rank*
1.	Before commencement of crop season	3.14	II	3.11	II	3.14	I	3.13	II
2.	During standing crop period	2.90	III	2.79	III	2.96	III	2.88	III
3.	During slack season	3.33	I	3.20	I	3.03	II	3.18	I

It is clear from table that organization of training programme during slack season was given first rank by both marginal and small farmers, whereas, it was moderately preferred by the medium farmers and was given IInd rank. Training before commencement of crop season the crop was given first rank by the medium farmers, whereas, it was assigned second rank by both the marginal and small farmers. Organization of training programme during standing crop period was given last rank by all the three categories of farmers.

Pooled figure indicated that organization of training programme during slack season was given Ist rank, whereas, training before commencement of crop season and during standing crop period was given 2nd and 3rd preference respectively by the farmers.

4.5.2 Duration of Training

Duration of training is an important component which also affect the training programme. Table 4.11 reveals that organization of one day training programme was given Ist rank by marginal farmers. However, it was given last rank by both the small and medium farmers. Two days training duration have been given Ist priority by both small and medium farmers whereas, it was given IInd rank by the marginal farmers.

Table 4.11: Farmers Option for Duration of Training

Sl. No.	*Duration for training*	*Marginal farmers (N=119)*		*Small farmers (N=54)*		*Medium farmers (N=27)*		*Pooled sample farmers (N=200)*	
		Mean score	*Rank*	*Mean score*	*Rank*	*Mean score*	*Rank*	*Mean score*	*Rank*
1.	One day	3.54	I	2.94	IV	2.33	IV	2.93	III
2.	Two day	3.37	II	3.70	I	2.92	I	3.33	I
3.	Three day	3.05	III	3.55	II	2.62	II	3.07	II
4.	One week	3.00	IV	3.09	III	2.37	III	2.82	IV

During of three days training was given 2nd rank by the small and medium farmers and it was ranked 3rd by marginal farmers. One week training programme was given last rank by the marginal farmer whereas, it was ranked third by both small and medium farmer. Pooled figure indicates that organization of training for two days was given Ist rank whereas, three days training was given 2nd rank. Duration of training for one day and one week were given third and fourth rank by the farmers.

4.5.3 Size of Trainees

The farmers wee asked to indicate their option regarding the appropriate size of training group. Their pattern was indicated in Table 4.12. Data presented in table shows that thee were similarities in opinion for different size of training groups by marginal, small and medium farmers. All the three categories of farmers ranked size of training group of 20 to 25 as the most appropriate. Size of group 26 to 50 was moderately appropriate and 51 to 75 farmers group was considered relatively less appropriate by all the three categories of farmers.

Table 4.12: Farmers Option for Size of Training

Sl. No.	*Number of trainees*	*Marginal farmers (N=119)*		*Small farmers (N=54)*		*Medium farmers (N=27)*		*Pooled sample farmers (N=200)*	
		Mean score	*Rank*	*Mean score*	*Rank*	*Mean score*	*Rank*	*Mean score*	*Rank*
1.	20 to 25	2.34	I	2.77	I	2.33	I	2.48	I
2.	26 to 50	2.23	II	2.70	II	2.26	II	2.39	II
3.	51 to 75	2.15	III	2.55	III	2.07	III	2.25	III

4.5.4 Opinion of Farmers in Respect of Training Methods

The farmers preference towards different methods by which they would like to be trained was also ascertained their opinion in respect of training methods have been depicted in Table 4.13.

Table 4.13: Farmers Option for Method of Training

Sl. No.	Duration for training	Marginal farmers (N=119)		Small farmers (N=54)		Medium farmers (N=27)		Pooled sample farmers (N=200)	
		Mean score	Rank	Mean score	Rank	Mean score	Rank	Mean score	Rank
1.	Lecture method	2.87	V	2.55	V	3.07	IV	2.83	IV
2.	Question answer method	3.54	I	3.11	II	3.18	II	3.27	II
3.	Practice trainees	2.58	VI	2.87	IV	2.74	V	2.73	V
4.	Audio-visual aids	2.99	III	2.92	III	3.11	III	3.00	III
5.	Demonstration	3.40	II	3.59	I	3.37	I	3.45	I
6.	Tour	2.15	VIII	2.07	VIII	2.37	VIII	2.19	VII
7.	Published material	2.98	IV	2.15	VI	2.51	VI	2.54	VI

Data presented in table revealed that in case of single method small and medium farmers gave their top priority in training methods as a demonstration at the field which was second preferred choice of marginal farmers. Question-answer method of training was ranked first by the marginal farmers which was second preferential choice of small and medium farmers. All the three categories of farmers assigned third rank to training with audio visual aids. Conducting tour as a method of training was the lowest rank by marginal, small and medium farmers. On the basis of pooled data, the rank position under single method was in the order of demonstration, question-answer method, audio-visual aid, lecture method, practice by trainees, published material and tour Ist, 2nd, 3rd, 4th, 5th, 6th, an d7th rank respectively.

4.5.5 Follow up Activities

Follow-up activities by the interaction of trained farmers with scientists/extension officer at regular interval was most preferred by all categories of farmers and was given Ist rank by marginal, small and medium farmers. It is evident that there were similarities in opinion of marginal and small farmers in the follow-up activities namely; by conducting demonstration in the farmer fields, supply of relevant literature at regular interval and maintaining a mailing list of trained farmers for periodical correspondence, supply of relevant literature and maintaining a mailing list of farmers were given 2nd, 3rd and 4th rank, respectively by both categories of farmers. The medium farmers had given 2nd, 3rd and 4th ranks to the maintaining the mailing list, by conducting demonstration in the farmers fields, and supply of relevant literature. Broadcasting talk through radio and television was given 5th rank by all the three categories of farmers as it was pressed at the bottom in order of preference.

The findings clearly revealed that all the three categories of farmers have similarity in their opinion in respect of size of training group, trainers categories and follow-up activities for organizing effective boro rice training programme. The marginal and small farmers showed similar preference with regard to venue of training in the village during the slack season. This might be due to the fact the farmers may not be able to leave their farms for longer period as it may hamper their farming operations. It is obvious from the finding that short duration training programme organized during the slack season at village level was most preferred by the maize growing farmers. In respect of venue for training, the medium farmers preferred the university/ college campus and their choicest time was before the commencement of crop seasons. Variation in the preference for marginal, small and medium farmers regarding venue, duration and time of training is a clear indication that nature of needs and desires of the farmers in quite complex.

Because of higher risk bearing capacity, better education and their being more change prone, the small and medium farmers liked the use of published materials, tour, lecture along with demonstration under combination of training methods. The findings also revealed that small and medium farmers showed similar preference with regard to use of training methods in different combinations. However, the choice of marginal farmers was the question-answer method in combination of practice by the farmers and lectures possibly because of their concern to acquire clear understanding and skill involved in the use of technology in shortest possible time. Because of high degree of commitment and providing right solution to the field problems by the scientists all the three categories of farmers liked interaction of trained farmers with scientists/extension officers at regular follow-up activities was conducting demonstration in the farmers field because of their concern to acquire clear understanding of technological skill involved in boro rice production method demonstration under the principle of "seeing in believing" and "learning by doing".

SECTION – VI

4.6 Training Strategy for Farmers

Indian agriculture during past six decades has passed through various phases and stages of development with growth rate of 2.5 to 3 per cent per annum. The agricultural production has increased four times (from 50 million tones in 1950 to more than 200 million tones recently) which has been possible not only due to horizontal (increase in cropping area well as cropping intensity) but vertical growth (cultivation of high yielding and fertilizer responsive varieties)in production. The vigorous efforts made in public as well as in private sector in the country for up-gradation of the skills and considerable change in attitude towards the production advance through planned approach ahs contributed much in the achievement.

However, a serious rethinking appears to be essential for human resources development specially at grass root level in farmers as well as extension personnel at village level for major break through in agricultural production. The demand of food and fodder is increasing day by day in light of ever growing population and area under cropping in decreasing. Hence the only way out left for increasing productivity and feeding the eve growing population is efficient utilization of natural production resources including modern or low cost technology for realization of the full potential of high yielding cultivars at field level. The situation demanded new extension approach for benefiting farming community specially marginal, small and medium cultivators. The up-gradation of their knowledge and skill in various components of production appears to be vital where government sponsored extension machinery or public sector may play a dominant role.

4.6.1 Training for Marginal Farmers

While analyzing the training needs of marginal farmers, wide technological gap was recorded in relation to improved seed and their treatment. There are indigenous sources of boro rice seed in the market having variation in production potential. The research results have exhibited more them 15% increase in yield due to the use of seeds of high yielding potential at the equal level of input management. The marginal farmers having limited resources do not want to bear risk and want to achieve maximum productivity at low input management and hence exhibited a wider gap in training need in this respect. The weed management is generally done manually and is not complicated and hence exhibited the lowest gap.

While analyzing the correlation of training need, significant correlation was recorded with farm size, cropping intensity, education, economic motivation and utilization of personal cosmopolites. Farmers with big farm size required more training for economic and efficient use of every part of land. The marginal farmers having small holding are interested harvesting more and more crops from the same piece of land and hence there is high positive correlation. Education of farmers created curiosity and sense of responsibility in use economic gain is the ultimate objective of any enterprise and its high correlation with training need is of vital importance.

The eagerness of the farmers to collect knowledge on personal basis and their use in their farming business for economic gain has also vital role in cultivation and this has high correlation with training need. Age of the farmer appeared to exhibit negative correlation. Younger the farmers they have more eagerness to acquire efficiency and skill which decreases with advancement in age thus decreasing the need of training.

5

Summary and Conclusion

Rice is globally a top ranking cereal human food, animal feed and as a source of large number of industrial products. It has acquired a dominant role in the farming sector and in the macro economy of the Asia and Pacific regions. In India, boro rice is emerging as an important traditional cereal crop, grown mainly for human food purposes, though recently its industrial uses are steadily increasing. It favorably responds to better crop management practices.

Boro rice has a high productivity potential to yield as high as 7-8.4 tonnes/ha and possibilities of further increase in productivity with better management. Bihar, has already substituted boro rice in several parts and has registered an impressive growth in rice production in recent years. The average yield of boro rice is 2-5 tonnes/ha in farmers fields, compared to the yields of 7-8.4 tonnes/ha in demonstration plots. One of the important reseason for yield gap is partial adoption of the production technology, and our inability to involve large number of farmers in agricultural development programme and to impart adequate knowledge and skills necessary for efficient use of ever changing agricultural technology as well as innovation. One of the ways to impart new knowledge and skills to the farmers is to have effective training programme for them.

The bright future for our agricultural production would depend on knowledge intensive and science based practices. The world never has been so vast a store house of knowledge in different fields as it has today. The emerging knowledge and technology have become the strategic resource for accelerating the process of advancement. The greatest and the most urgent need of the day is to seize the opportunities they offer. This signifies the paramount importance of speedy and effective transfer of available knowledge and technology to the users. It aims in helping each individual to reach his maximum potential by way of increased knowledge, improved skills and changed attitudes enabling him to perform his job.

Training builds up the professional competency of the farmers and extension personnel too. The farmers of low socio-economic status with less education and having less knowledge about the various component of winter boro rice technology can also be well equipped through training programmes. The training needs of farmers and extension personnel should be determined in clear terms to provide proper direction and thrust to training efforts. It is a undisputed fact that raining needs of farmers are presently not being given due attention. So, it is important first to assess as to how far the existing programmes are catering to their specific needs with respect to the adoption of technological innovation in agriculture. The present study "Assessment of Training Needs in Designing Strategy for Increasing Boro Rice

Productivity in Madhubani District of North Bihar" had planned to arm the scientists and extension workers with the power of knowledge about the training needs of different categories of farmers and to develop a sound training strategy to bridge the gap between the boro rice production technology available in the research station and farmers practice in the fields.

5.1 It was Imperative to Conduct this Study with the following Objectives:

5.1.1 To identify agro-economic, socio-psychological and communicational characteristics of boro rice farmers.

5.1.2 To analyze the extent of technological gaps in adoption of scientific boro rice production technology.

5.1.3 To assess the training needs in respect of scientific boro rice production as perceived by farmers and extension personnel.

5.1.4 To find out the association of the training needs with certain selected agro-economic, socio-psychological and communicational variables.

5.1.5 To seek the opinion of farmers in respect of different components of training programme.

5.1.6 To develop training strategy for different categories of farmers for increasing boro rice production.

The proposed study was purposely conducted in the Madhubani district of North Bihar occupying the highest area and production of boro rice in Agro-climatic Zone

II. Two blocks i.e. Madhepur and Lakhnaur were purposively selected for this study.

For the selection of respondents, a comprehensive list of different categories of farmers were prepared in each sample block. The farmers of the selected blocks were classified into three categories i.e. marginal, small and medium. A proportionate random sample was drawn from each of the three categories of farmers. In this way 119 marginal, 54 small and 27 medium farmers (total 200 respondents) were selected, which formed the sample of the study. Based on an extensive review of literature, discussion with the experts and pilot study, a manageable system of important variables likely to be related with training need of the farmers were selected.

Training need is the dependent variable for the present study, which was measured through a three point rating scale and average choice score as suggested by Singh (1980). For the selection of main areas and sub-areas of training components in respect of boro rice technology a comprehensive list of main areas and sub-areas of training priority was prepared. Then it was discussed with scientist, and other extension functionaries and finally after pre-testing the structure schedule in the area of study, the final training area and sub-area were selected. A separate schedule was prepared to obtain the views of extension personnel in respect of differential training need perception for the three categories of farmers. The same method was adopted for analyzing the training need priority for different categories of farmers as perceived by extension personnel.

5.2 The Independent Variables

The variables selected for the study were grouped as follows:

5.2.1 Agro-Economic

Farm size, irrigation index, cropping intensity, farm mechanization and technological gap.

5.2.2 Socio-Psychological

Age of farmer, education of farmer, family size, knowledge level, risk orientation, economic motivation and market orientation.

5.2.3 Communicational

Utilization of mass media and utilization of personnel cosmopolite.

For measurement of independent variables tests were developed and some of the scales developed by other research workers were used. To know the opinion of farmers in respect of different components of training, a four point rating scale was used. The data were collected by the researcher himself by interviewing the respondents through a schedule specially constructed for the purpose of the present investigation.

5.3 The Important Findings of the Study are:

5.3.1 Differential Characteristics of Marginal, Small and Medium Farmers with Respect to Agro-Economic, Socio-Physchological and Communicational Variables.

5.3.1.1 In comparison to both the small and medium farmers, the marginal farmers had significantly smaller farm size, lower irrigation index, lower farm mechanization, lesser knowledge and less utilization of personal cosmopolite sources of information.

5.3.1.2 The marginal farmers had significantly more technological gap in recommended of practices and also more market orientation, in comparison to both the small and medium farmers, However, the marginal farmers had significantly lower educational level and higher age than the medium farmers.

5.3.1.3 The small farmers had significantly lower cropping intensity, more technological gap and higher economic motivation in comparison to medium farmers.

5.3.1.4 No significant differences was found between the small and medium farmers with respect to irrigation index, farm mechanization, age of the farmers, knowledge level, risk orientation, market orientation utilization of mass medium and utilization of personal cosmopolite sources of information. There was no significant difference between the marginal, small and medium farmers with respect to family size.

5.3.2 *Extent to Technological Gap in Boro Rice Cultivation.*

5.3.2.1 In case of marginal farmers, the highest technological gap was observed in the practice of improve seed and their treatment followed by fertilizer application, sowing methods and irrigation in order of preference by the farmers.

5.3.2.2 The highest gap was found in the practice of "Fertilizer application" by small as well as medium farmers. It was interesting to observe that least technological gap was also similar by both small and medium farmers in the practice of post harvest technology.

5.3.3 Training Need Perception of Farmers in Boro Rice Technology

5.3.3.1 Improved cultural practices emerged as the top most priority training area as there was common consciousness among the three categories of farmers who preferred this training area as most needed.

5.3.3.2 The findings reveals that use of HYVs, integrated nutrient management and water management were preferred as a seasonal priority area by the marginal, small and medium farmers respectively.

5.3.3.3 Small and medium farmers have shown similar priority in weed control and land preparation whereas, water management and integrated pest management were accorded similar priority by medium & small farmers.

5.3.3.4 In the sub-areas of training, reduced tillage, transplanting of rice seedling, identification of pest, time of irrigation, method of storage and weed control by manual, chemical and mechanical method have been given top priority by all the three categories of farmers.

5.3.3.5 The findings reveals that small and medium farmers have shown similar priority by assigning same priority in the different sub-areas of use of

HYVs, land preparation, integrated nutrient management, weed control, cropping system, water management and post harvest technology.

5.3.4 Training Need Perception by Extension Personnel

5.3.4.1 Use of HYVs was perceived by the extension personnel as the most needed area for marginal farmers followed by improved cultural practices and integrated nutrient management in order of preference.

5.3.4.2 The finding reveals that extension personnel accorded 1st, 2nd and 3rd place for integrated nutrient management, improved cultural practices and integrated pest management respectively for both small and medium category farmers.

5.3.4.3 There was strong degree of agreement among rank pattern of extension personnel v/s medium farmers followed by extension personnel v/s marginal and small farmers.

5.3.5 Association of Variables with Training Need in Boro Rice Technology

5.3.5.1 The variable cropping intensity and economic motivation were found to be positively and significantly associated with training need in all the three categories for farmers.

5.3.5.2 The variables farm size; education of the farmer and utilization of personal cosmopolite were positively and significantly associated with training need of marginal farmers in boro rice technology.

5.3.5.3 Technological gap and utilization of mass media were found to be positively and significantly associated with training need of small farmers.

5.3.5.4 Variable farm mechanization, risk orientation, market orientation, utilization of mass media and utilization of personal cosmopolite sources were found to have positive and significant association with training need of medium farmers.

5.3.5.5 The variable age was negatively and significantly associated with training need of all the three categories of farmers. Whereas family size was negatively and significantly associated with training need of small farmers.

5.3.5.6 The variable farm size, cropping intensity, education of the farmer, risk orientation, economic motivation utilization of mass media and utilization of personal cosmopolite were found to have positive and significant association with the training need of farmers in general (pooled).

5.3.6 Prediction Potency of Variables towards the Training Need in Boro Rice

5.3.6.1 Variables farm size, education motivation and cropping intensity emerged as significant contributors towards identifying the training need of marginal farmers and were ranked 1st, 2nd, 3rd and 4th respectively.

5.3.6.2 Only three variables namely economic motivations, cropping intensity and utilization of mass media emerged as powerful predictor in case of small farmers in boro rice technology and were ranked 1st, 2nd, 3rd respectively.

5.3.6.3 Variable market orientation, utilization of mass medium and economic motivation emerged as power predictors in case of medium farmers in boro rice technology and were ranked 1st, 2nd, 3rd respectively.

5.3.6.4 Cropping intensity, economic motivation, risk orientation & irrigation index emerged as significant contributes towards training need of all categories of farmers (pooled) in boro rice and were ranked 1st, 2nd, 3rd and 4th respectively.

5.3.7 Direct and Indirect Effect of Independent Variables on Training Need of the Farmers.

5.3.7.1 The variable farm size, economic motivation and market orientation were deciding factors affecting the training need as these variable had the highest substantial direct effect on training need of marginal, small and medium farmers respectively.

5.3.7.2 The variable cropping intensity, utilization of mass media and utilization of personal cosmopolite were found to be important factors affecting the training need as these variables had maximum substantial indirect effect on training need of marginal, small and medium farmers respectively.

5.3.7.3 The finding reveals that substantial indirect effect mostly channeled through the variables farm size, age of the farmers, education of the farmers and economic motivation in case of marginal farmers.

5.3.7.4 In case of small farmers the substantial indirect effect mostly channeled through the variables cropping intensity, age of the farmer, economic motivation and utilization of mass media.

5.3.7.5 In vase of medium farmers the substantial indirect effect mostly channeled through the variables risk orientation, market orientation, utilization of mass media and utilization of personnel cosmopolite sources of information.

5.3.8 Opinion of Farmers in Different Components of Training Programme

5.3.8.1 Venue, Time, Duration and Size of Training

5.3.8.1.1 All the three categories of farmers showed their desire for organizing training programme with a group of 20 to 25 farmers to be imparted by rice scientist and also their interaction with extension officer and worker at regular intervals after participating in training programme.

5.3.8.1.2 Organization of training programme in the village during slack season was most preferred by small farmers. However, medium farmers preferred it in the university/college campus and time of training before commencement of crop season.

5.3.8.2 Training Methods

5.3.2.1 In case of single method of training question-answer method was ranked first by the marginal farmers whereas small and medium farmers gave top priority in training method as demonstration at the field.

5.3.2.2 Under combination of two methods, lecture + practice by trainees was ranked first by marginal farmers whereas small and medium farmers gave top priority in training methods as lecture + method demonstration.

5.3.2.3 Under combination of three methods, lecture + method demonstration + practice by trainees was ranked first by marginal farmers whereas small and medium farmers gave top priority in training methods as lecture + method demonstration + published material.

5.3.2.4 Under combination of four methods lecture + method demonstration + practice by trainees + published material was ranked first by marginal farmers whereas small and medium farmers gave top priority in training methods as method demonstration + question-answer + tour + published material.

5.3.2.5 Under combination of five methods A.V. Aids + practice by trainees + question-answer + published material + tour was ranked first by marginal farmer. Whereas small and medium farmers gave top priority in training methods as lecture + method demonstration + practice by trainees + tour + publication.

5.4 Implication of the Study

The findings of the present study may be meaningful for various institutions and extension personnel to prepare suitable strategy for organizing the training programmes for different categories of farmers specially in boro rice. Based on the findings, some of the implications for farmers participation in development of boro rice programme are as follows.

The findings provide the extent of technological gap in the adoption of various components of production technology by the boro rice growing farmers. It is necessary to reduce the technological gap for increasing the crop yield to desired level through appropriate training in the new innovation.

The findings of the study provide perception of training need by different categories of farmers in the main areas and sub-areas of boro rice production, which needed to be looked into while organizing need based and client oriented training programme. In addition, it also provides the perception about the training need by extension personnel for different categories of farmers.

The finding of projects training need in improved cultural practices in the field for better establishment & congenial growth condition for crop plants as the most needed area irrespective of the categories of the farmers. The scientist & extension personnel engaged in technology development & technology transfer approach were required to give emphasis on "no cost and low cost technology" for boro rice production, so that more benefit may accrue to the farmer-client. This reflected that farmers are interested in knowing about prerequisites for profitable production which can be achieved by optimal eco-system management. This seems to be a logical & innovative attempt for making training programmes more purposeful effective.

It has also emanated through the research study that training need in all the categories of farmers are influenced by a set of selected agro-economic, socio-psychological and extension communicational variables. Utilization of personal cosmopolite sources, cropping intensity & economic motivation which exhibited significant association with training need irrespective of categories of farmers required due attention while preparing the training strategy for boro rice cultivation. Age of the boro rice growing farmers displayed significant but negative association with the training need in all the three categories of farmers which clearly indicated that more was the age of the farmers lesser was the incentive for training need. This needed to be kept in mind while formulating training strategy for the farmers engaged in boro rice cultivation.

The study also highlights the differential choice for different training components for organizing effective training programmes for the farmers. On the basis of findings it can be inferred that training for the farmers should be organized at the village level during slack season preferably for two days duration with a size of 20 to 25 trainees in numbers. The findings reveal that there is a general perception for imparting training by combination of different training methods like method demonstration, published material, practice by the trainees and educational tour. Instead of single method as generally practiced by the training organizer.

There is a need to improve the information stock of the farmers in different agro-climatic regions. In fact, the extension service network needs to be strengthened in all regions. It should also needs to be fully supported by a strong service sector, which would ensure timely supply of required inputs to increase agricultural productivity. The agricultural planning process in the 21st century has to take full advantage of the rich and enormous potential of information technology for rapid information dissemination in the rural areas.

It is high time that India takes stock of all available technologies and devises ways of using them for food grains and other corp production. This not only calls for extensive exploration of alternative possibilities but also requires an immediate long term technology dissemination for Indian agriculture. It may be reiterated that our country has great expectations from its agricultural scientists and the system which is responsible for transferring agricultural technologies to the farmers. It this direction the findings of present investigation have yielded the clues for the effective transfer of technology and its adoption by the farmer clients within the limitation of the study. This may help in exploring the possibilities for rapid transfer of technology with needed modification according to the socio-cultural and agro-economic selling of the farmers in different regions.

5.5 Suggestion for the Future Work

Based upon the present investigation the following are a few projections, which should be considered for future study.

5.5.1 The present study was conducted with a sample of 200 farmers of Two Blocks. In order to have greater generalization of findings such study should be conducted with a large sample of farmers & field functionaries in different agro-climatic zones of the state.

5.5.2 Studies on impact of training on adoption behaviour of farmers should be planned and conducted.

5.5.3 Study on stand using methodology for farmers participant technology development & technology transfer should be planned and conducted.

5.5.4 Studies on effectiveness of various combinations of training for farmers should be planned.

5.5.5 Studies on technological gap and constraints analysis with respect to boro rice production technology should be conducted.

Appendix I

Schedule for Interviewing Farmers on Training Needs for Different Category of Farmers in Boro Rice Production Technology

(1) Name

(2) Place

(3) Designation

(4) Qualification

(5) Working experience

The different items of package of practices of winter maize technology are given in table. Please mention your preferential choice of training in different main areas for all the three categories of farmers. Your judgment should be based on your estimate of the need of knowledge about items for a farmers and their existing level of knowledge in each items for the crop.

Sl. No.	*Main areas of training*	*Response*								
		Marginal farmers			*Small farmer*			*Medium farmer*		
		1st choice	*2nd choice*	*3rd choice*	*1st choice*	*2nd choice*	*3rd choice*	*1st choice*	*2nd choice*	*3rd choice*
1.	Land preparation									
2.	Use of H.Y.Vs									
3.	Seed treatment									
4.	Improved cultural practices									
5.	Integrated nutrient management									
6.	Weed management									
7.	Cropping system									
8.	Integrated pest management									
9.	Water management									
10.	Post harvest technology									

Bibliography

Aldrich, Daniel G. (1988). *Educational and Training Needs to meet the Agricultural Production Requirement of the 21st Century.* In Raman et al. (1988) (Q.V.), pp. 179-184.

Anonymous (1984-85). *Monitoring-cum-Evaluation Survey Report,* Department of Agriculture, Bihar (Kharif, 1983-94).

Barroah (1965). *Farmer's training in India,* Kurukshetra, 14:12.

Bhattacharya, S.P., Das, D.; Malti, S.; Ghosh, B. and Ray, M. (2000). *Bio-efficacy of quindorac 15 SC in controlling weeds of transplanted boro rice.* Environment and Ecology. 1893): 557-560.

Biharamkar, M.S. et al. (1997). *"Training needs of master trainer in T and V system"* Maharastra Jour of Ext. Edn. Vol. XVI: 47-52.

Chang, T.T., G.C. Loresto and O. Tagumpay (1972). *Agronomic and growth characteristics of upland and lowland rices.* In Rice Breeding, pp. 645-664.

International Rice Research Institute, Los Banos, Philippines.

Chaturvedi, T.N. (1979). *Institution Building for Training,* in Mathur et al. (Eds.) (1979) (q.v.), Chapter 12: pp. 141-150.

Chaudhar, S.K. (1999). *Training need of farmers if relation to high yielding rapeseed andmustard in adopted village of Muraul Block (Muzaffarpur district).* Unpublished M.Sc.(Ag.) Thesis, Department of Extension Education, RAU, Bihar, Pusa, Samstipur.

Chung, G.S. (1979). *Report of rice cold tolerance workshop.* Int. Rice Res. Inst. Lao Banos, Philippines, 7-19.

Coly, A. (1980). *Testing rice varieties of indica type for cold tolerance at seedling stage.* Japan 1 Crop Scie. 49(4): 543-547.

Dahama, O.P. (1973). *Extension and rural welfare,* Ram Prasad and Sons, Agra-3.

Extension Education Institute (1971). News from APAU (Rajendra Nagar) 2 : (9) : 4.

Gaikwad, V.R. (1988). *Training of Agricultural Administrators, Second Generation Challenges,* Indian Journal of Public Administration, 34 (3) : 690-702. Special Number on Training in Public Administration: The changing perspective, July-September.

Geues, J.D. (1964). *Training farmers at extension service short courses,* Extension in the near earst (5) : 13.

Ghuman, P.; Hansra, B.S. and Mehta, A.K. (1999). *"Training needs of farm women in a agriculture"*, Jour. Ext. Edn., 10 (2) : 2359-69.

Gogoi, Nitumoni; Baruah, K.K. and Gogoi, N. (2000). *Effect of cold hardening and gibberellic acid on growth and yield of boro rice*. Indian Journal of Plant physiology, 5(4): 339-343.

Gupta, M.P. (1982). *Training needs of farmers in Himachal Pradesh*. Indian J. Extn. Edn., 18 (3&4): 70-71.

Hall, Milton (1962). *Definition of employee training C.F. Role of trainee in extension training centres*. (T. Somasundram). Kurukshetra 10(10) : 12-15.

International Rice Research Institute (IRRI) (1975). *Major research in upland rice*. Los Banos, Philippines.

Islam, Z. (1996). *Yellow stem borer : a threat to boro rice in the coastal belt of Bangladesh*. Bangladesh Journal of Entomology. 6(1-2): 45-52.

Kalita, H.K. and Sarmah, R.C. (1999). *"Training needs of village level extension workers"*. Jour. Etn. Edn., 10(1) : 2281-85.

Kaw, R.N. (1995). *Analaysis of divergence in some cold tolerant rices*. Indian J. Genet. 55(1): 84-89.

Kaw, R.N. and Khush, G.S. (1985). *Heterosis in traits related to low temperature tolerance in rice*. Philipp. J. Crop Sci. 10: 93-105.

Khemmani, Dr. M. (1983). *Turning Thoughts Towards Training, Manila, Philippines, Asia and Pacific Programme Planning (DTCP),* United Nations Development Programme (UNDP) Training Tips No.1 Editor: Andrew P. Bartlett TT 601.

Kubr, Milan and Joseph Prokiopenko (1989). *Diagnosing Management Training and Development Needs : Concept and Techniques*, Geneva, International Labour Office, Management Development Series No.27.

Kumar, Navin; Thakur, R.B.; Singh, R.; Pandey, S.K. and Kumar, N. (1998). *Influence of level and time of nitrogen application on the incidence of major pests of rice in boro season*. Shashpa. 5(1): 73-76.

Kumari Shushma, N.P. and Bhaskaran, C. (1995). *"Assessment of Training needs of farmers in agriculture"*. Journal of Tropical Agriculture, 33 : 1, 59-61.

Lynton, R.P. and Pareek, Udai (1967). *Training for Development Homewood* (Illinoid). Richard D. Irwin Inc. and the Dorsey Press : pp. 3-28.

Lynton, Rolf P. and Udai Pareek (1990). *"Training for development"* New Delhi, Vistaar Publications, Second Edition.

Mahmood, R. (1997). *Imapct of air temperature variations on the boro rice phenology in Bangladesh: implication for irrigation requirements*. Agricultural and Forest Meteorology. 84(3-4): 233-242.

Mathiyazhagan, T.; and Singh, R.P. (1986). *Training needs of banana growers*, IJEE Vol.22 No.1 & 2.

Mehta, S. and Malaviya, A. (1997). *"Agricultural training needs of farm women of Haryana"*. Journal of Dairying, Foods and Home Sciences, 16 : 1, 13-22.

Mishra, D.C. (1990). *New Directions in Extension Training*, Published by Directorate of Extension, New Delhi. pp. 9-24.

Molla, A.H.; Quayyum, H.A. and Islam, M.S. (1996). *Comparative physiology of boro rice seedling under upland and lowland seedbeds*. Bangaldesh Journal of Botany. 25(2): 243-245.

Morison, James, H. (1976). *Determining Training Needs,* incraig (Ed.) (1976), Chapter 9 : pp. 9-1 to 9-77.

Narshimha, N. and Rao, M.S.K. (1983). *Impact of training on knowledge and adoption,* Kurukshetra, XXXI (20) : 12-14.

Patil, S.O. and Kale, J.V. (1972). *Vocational training needs of farmers with special reference to the contents and type of training.* Ind. J. Extn. Edn., 8(3&4) : 10-24.

Paul, Samuel, John, C. Ickis and Jacob, Levitsky (1989). *Educating Managers for Business and Government : A review of International Experience,* Washington, D.C. The World Bank discussion papers 54 : June.

Prasad, C. (1983). *"Training for Agriculture".* Ind. Jour. Extn. Edn., Vol; 19, No. 1&2.

Prasad, M.V. & Mantyanyum, N. (1992). *Training needs of tribal farmers on paddy cultivation.* Indian Jour. Extn. Edn. Vol. 28 No. 1&2 (1992).

Pushpa, J. and Seetharaman, R.N. (1999). *"Training needs of farmers on integrated farming".* Jour. Etn. Edn., 10(1). 2293-98.

Ram, Ganga and Mathur, P.N. (1981). *Training priorities in Krishi Vigyan Kendra.* Indian J. Extn. Edn., 17 (1&2): 48-53.

Ramegowda, B.L.; B.T. Chikkaputtee Gowda and M.K. Sethu Rao (1989). *Training Needs of Agricultural Change Agents for Communication Methods,* Maharastra Journal of Extension Education, 8: 291-93.

Rao, M.K. Sethu (1975). *Application of Lynton-Pareek model in farmer's training.* Indian J. Extn. Edn., 11 (1&2), 76-80.

Raut, R.S. et al. (1997). *"Training need & expectation of farm youth in cultivation of dry lend horticultural corps"*. Maharastra Journal of Extension Education. Vol. XVI: 53-57.

Ravichandran, V.; Muthammal, K. and Krishnan, J. (2000). *"Training needs of farm women in rice cultivation"*. Jour. Etn. Edn., 11(1): 2662-65.

Sailaja, A. and Reddy, M.N. (1999). *"Training needs of farm women in paddy and additional activities"*. Jour. Extn. Edn., 10(2): 2419-24.

Samba (1961). *Training the Gram Sahayaks*. Kurukshetra Chapter 28. A symposium on community Development in India, Delhi-6. The publication division. Ministry of Information and Broadcasting, Govt. of India.

Sander's, E.N. (1967). *Importance of extension training in agricultural production report of all India seminar on extension research,* ISEE, division of Agricultural Extension, IARI, New Delhi.

Satyanarayan, I. and Bhaskaran, K. (1971). *Training needs of farmers in Hyderabad district*. Extension Trainer, 1(1): 4-47.

Schein, Edger H. (1980). *"Organizational Psychology"*, Englewood Cliffs, N.J., Prentice Hall, Third Edition.

Shahi, B.B. and Khush, G.S. (1985) *analysis and cold tolerance in rice ge Rice* Res. Inst. (IRRI): 429-435.

Sharma, K.A.L.N. & Murthy, A.S. (1971). *Training needs as perceived by farmers,* Kurukshetra, 19(15) : 7.

Sharma, R.D. (1987). *The Story of Rice Banos,* Philippines.

Shete, N.B. (1978). *Agricultural training needs of tribal farmers in Maharashtra*. Ind. J. Extn. Edn., 14(3&4): 65-67.

Shibata, M.; Sasaki, K. and Shimazaki, *Effect of rice temperature and water tem each stage of growth of lowland rice. Temperature and water departure on the of sterile grain.* Prod. Crop Sci. Japan. 395-408.

Siddiq, E.A. (1996). *Export of rice and products: Prospects and problem Rice India,* p.8-11.

Siddiq, E.A. (1996). *Rice Development opportunities*. The Hindu Survey of Agriculture.

Virmani, B.R. (1989). *Management Training and Development An Evaluation Approach,* Indian Journal of Training and Development, XIX (5) 5-517, September-October.

Yoshida, S. (1981). *Fundamentals of rice crop science*. The International Rice Research Institute, Los Banos, Philippines.

Yuan, L.P.; Yang, Z.Y. and Yang, J.B. (1994). *Hybrid rice in China-In "Hybrid rice Technology,* New Developmental and Furture Prospects (S.S. Winami, ed.), pp. 143-147. Int. Rice Res. Institute, Manila, Phlippines.

Zaman, S.K.; Razzaque, M.A.; Karim, S.M.R. and Bhuiyan, NI.I. (1997). *Rice response to phosphorus in wet land soil*. Pakistan Journal of Scientific and Industrial Research. 38(11-12): 438-440.

